AF685048

Radiation Effects of Advanced Electronic Devices and Circuits, 2nd Edition

Radiation Effects of Advanced Electronic Devices and Circuits, 2nd Edition

Guest Editors

Chang Cai
Yaqing Chi
Li Cai

MDPI

Basel • Beijing • Wuhan • Barcelona • Belgrade • Novi Sad • Cluj • Manchester

Guest Editors

Chang Cai
School of Physics and Materials Science
Guangzhou University
Guangzhou
China

Yaqing Chi
College of Computer
National University of Defense Technology
Changsha
China

Li Cai
Institute of Modern Physics
Chinese Academy of Sciences
Lanzhou
China

Editorial Office
MDPI AG
Grosspeteranlage 5
4052 Basel, Switzerland

This is a reprint of the Special Issue, published open access by the journal *Electronics* (ISSN 2079-9292), freely accessible at: https://www.mdpi.com/journal/electronics/special_issues/YC9AGK4QG5.

For citation purposes, cite each article independently as indicated on the article page online and as indicated below:

Lastname, A.A.; Lastname, B.B. Article Title. *Journal Name* **Year**, *Volume Number*, Page Range.

ISBN 978-3-7258-4833-1 (Hbk)
ISBN 978-3-7258-4834-8 (PDF)
https://doi.org/10.3390/books978-3-7258-4834-8

© 2025 by the authors. Articles in this book are Open Access and distributed under the Creative Commons Attribution (CC BY) license. The book as a whole is distributed by MDPI under the terms and conditions of the Creative Commons Attribution-NonCommercial-NoDerivs (CC BY-NC-ND) license (https://creativecommons.org/licenses/by-nc-nd/4.0/).

Contents

About the Editors

Chang Cai

Chang Cai received his Ph.D. degree from the Institute of Modern Physics, Chinese Academy of Sciences, in 2021. After graduating, he joined Fudan University as a researcher. In 2024, he began working at Guangzhou University, where he continues to make significant contributions to the field of radiation effects in integrated circuits. Dr. Cai's research focuses on a wide range of topics within this field, including the radiation mechanisms for advanced devices. His work also includes the design of fault-tolerant Field-Programmable Gate Arrays (FPGA) and Systems on Chip (SoC), where he develops strategies to enhance radiation hardness through both layout and logical design. In addition to his theoretical and design work, Dr. Cai is actively involved in practical experiments. He conducts radiation experiments in extreme conditions to test and validate the robustness of various electronic devices. His expertise includes single-event effect characterization and evaluation techniques for VLSI, which are essential for ensuring the reliability of complex integrated circuits in radiation environments. Dr. Cai has published more than 30 peer-reviewed papers as the first or corresponding author in recent years. His research has been widely recognized and presented at numerous international conferences, including prestigious symposiums such as the IEEE IRPS.

Yaqing Chi

Yaqing Chi, Ph.D., is an associate researcher and master's supervisor at the School of Computer, National University of Defense Technology, director of the Hunan Aerospace Society, and member of the "High Performance Microprocessor Technology" Innovation Team of the Ministry of Education of China. He received his Ph.D. degree in 2009 from the National University of Defense Technology and then worked there up until now. He leads a team focusing on radiation effect testing for advanced integrated circuits, and he has accomplished many key technological advancements such as single-event effect characterization and the measurement for nano-integrated circuits, radiation resistance testing and the evaluation for complex microprocessors, and so on. He participated in the development and evaluation of many aerospace microprocessors, which have been widely applied in aerospace projects such as satellites, rockets, spacecrafts, space stations, and deep space explorations. His research interests include radiation effects in integrated circuits, radiation hardening techniques through the design for microprocessors based on advanced semiconductor processes, and high-speed radiation-tolerant SerDes and FPGAs. He has won an award of national scientific and technological progress and four provincial and ministerial level awards, obtained more than 20 national invention patent authorizations, and published more than 40 papers in important journals and conferences in the field of integrated circuit radiation effects.

Li Cai

Li Cai, Ph.D., researcher, had been selected for the CAS Talent Program Youth Project and has served as a doctorate supervisor at the Institute of Modern Physics, Chinese Academy of Sciences. After graduating with a master's degree in 2009, she worked for the Chinese Academy of Atomic Energy (CIAE) and obtained her doctoral degree from the CIAE in 2017. She began to work at the Institute of Modern Physics in 2020. Her primary research field is in the radiation effects of semiconductor devices, especially experimental accelerator-based technology. Dr. Cai has experience with broadbeam and microbeam irradiation experiments, and she has established a sample temperature measurement and control system based on the SEE test terminal, studying the impact of temperature on SEE in SRAMs and inverter chains. During her work, she has collaborated with several universities and institutes to study radiation effects on SRAMs, DSPs, FinFET inverter chains, CMOS image sensors, MRAM, etc. She has won a second prize for National Defense Science and Technology Progress and the first prize for the China National Nuclear Corporation Science and Technology Award, and she has published more than 20 papers as the first or corresponding author in important journals and conferences in the field of radiation effects on semiconductor devices. As a project leader, she has led multiple scientific research projects, including projects affiliated with the National Natural Science Foundation of China, national defense pre-research, talent projects, etc.

electronics

MDPI

Editorial

Radiation Effects of Advanced Electronic Devices and Circuits, 2nd Edition

Chang Cai [1], Yaqing Chi [2,3,*] and Li Cai [4,*]

1 School of Physics and Materials Science, Guangzhou University, Guangzhou 510006, China; caichang@gzhu.edu.cn
2 College of Computer, National University of Defense Technology, Changsha 410073, China
3 Key Laboratory of Advanced Microprocessor Chips and Systems, National University of Defense Technology, Changsha 410073, China
4 Institute of Modern Physics, Chinese Academy of Sciences, Lanzhou 730000, China
* Correspondence: yqchi@nudt.edu.cn (Y.C.); caili@impcas.ac.cn (L.C.)

1. Introduction

In recent years, the expanding use of advanced electronics in harsh radiation environments has heightened reliability demands, prompting extensive research into radiation-effect modeling and advanced radiation hardening design methodologies that serve these applications [1–6]. The relentless scaling of integrated circuit (IC) fabrication processes and the growing complexity of electronic devices have further heightened their susceptibility to radiation effects, presenting clear yet compelling challenges that will continue to drive future research [7–16]. The second edition of "Radiation Effects of Advanced Electronic Devices and Circuits" features thirteen high-quality contributions that showcase emerging applications and address recent breakthroughs (Contributions 1–13). These contributions collectively advance radiation hardening and reliability assessment across critical domains. Novel hardening strategies enhance fault tolerance in aerospace computing systems, moving beyond conventional monolithic approaches through hierarchical techniques. Significant progress is reported in understanding fundamental radiation damage mechanisms within wide-bandgap semiconductor platforms, covering complex interactions involving Total Ionizing Dose (TID), displacement damage, and Single-Event Effects (SEEs). Research advances the characterization of synergistic effects where multiple degradation mechanisms concurrently impact device lifetime and performance. Pioneering methodologies are established for mission-specific reliability validation, enabling more accurate prediction of failure pathways under representative radiation environments. Critical work addresses electromagnetic compatibility challenges through innovative shielding design principles and predictive modeling techniques. Studies also decode transient radiation damage evolution in energy conversion components and establish operational safety frameworks for power electronics in extreme environments. This Special Issue significantly strengthens the foundational knowledge for developing radiation-tolerant electronics, providing essential insights into physics mechanisms while delivering practical hardening solutions and assessment frameworks for increasingly demanding aerospace and terrestrial applications.

2. Highlighting Key Contributions

The thirteen articles comprising the second edition of "Radiation Effects of Advanced Electronic Devices and Circuits" synthesize cutting-edge developments in radiation effects, showcasing the latest research advances across critical semiconductor technologies. These

 https://doi.org/10.3390/electronics14142896

encompass CNN accelerators for aerospace computing, Gallium Nitride (GaN) High Electron Mobility Transistors (HEMTs), high-barrier beta-Gallium Oxide (β-Ga_2O_3) Schottky Barrier Diodes (SBDs), silicon MOSFETs, and Indium Phosphide Heterojunction Bipolar Transistors (InP HBTs), as detailed in Contributions 1–13.

Integrating machine learning into space missions such as remote sensing imaging and space debris management has become a rising trend. Employing Convolutional Neural Network (CNN)-based edge computing in remote sensing tasks enables real-time processing of captured images, lowers costs, and extends mission duration and battery life. However, deploying CNN accelerators in aerospace computing faces severe threats from Single Event Upset (SEU)-induced weight bit flip errors. Cai, Y. et al. (Contribution 1) merges weight-limiting hardening techniques with a Triple Modular Redundancy (TMR) architecture. The SEU vulnerable layers are observed using the fault injection method. It develops hierarchical hardening strategies that achieve impressive gains in fault tolerance over conventional monolithic approaches.

The unique properties of wide-bandgap semiconductors underlie their diverse applications in fields such as power electronics, RF electronics, and optoelectronics. As a strong contender for fourth-generation semiconductors, research on beta-Gallium Oxide (β-Ga_2O_3) is shifting from fundamental material properties to device applications. Fu, W., et al. (Contribution 2) decouples the contribution of temperature to TID effects in β-Ga_2O_3 SBDs, uncovering high-temperature annealing effects while simulating complex real-world operational environments. Zhao, P. et al. (Contribution 3) comprehensively evaluates the TID and displacement damage effects induced by X-ray and 1 MeV reactor neutron irradiation on β-Ga_2O_3 SBDs, clarifying post-irradiation performance and physical damage mechanisms. In addition, GaN HEMTs, as cornerstone devices of third-generation semiconductors, play vital roles in radiation-tolerant electronics. Huang, H. et al. (Contribution 4) deciphers the cascading failure pathways in GaN HEMTs by proton and heavy-ion irradiation and quantitatively maps the degradation hierarchy from interface trap generation and 2D electron gas depletion to drain leakage runaway. Moreover, InP semiconductors are widely used in aerospace applications due to their superior material properties, yet their susceptibility to space radiation requires critical attention. Zhao, X., et al. (Contribution 5) establishes a degradation model for InP HBTs during proton irradiation and annealing processes, providing deep insights into proton-induced damage mechanisms. This study pioneers a comprehensive research framework integrating both irradiation and annealing effects.

Expanded use of advanced integrated circuits in aerospace systems elevates the significance of MOSFET reliability analysis. He, Y., et al. (Contribution 6) systematically investigates the synergistic mechanisms of TID and hot carrier injection on NMOS device performance degradation, revealing their combined influence in lifetime assessment and filling a critical gap in quantitative analysis of combined effects. As MOSFETs constitute fundamental building blocks of memory cells, their radiation susceptibility directly impacts SRAM reliability. Therefore, Zheng, S. et al. (Contribution 7) examines temperature-dependent SEU cross-section variations in 28 nm SRAM, pioneering atmospheric neutron irradiation studies for modern-process SRAM. This work addresses a research void while elucidating the physical mechanisms underlying temperature-dependent SEU cross-section changes.

With the trend of modern electronic devices toward high-speed operation, high integration, and the wide adoption of wireless communication technologies, electromagnetic interference (EMI) and electromagnetic compatibility (EMC) issues have become increasingly critical. Consequently, efficient electromagnetic shielding technology has emerged as a key solution to ensure reliable operation of electronic equipment. Kwon, J.H., et al. (Contribution 8) proposed a structure integrating a dual-layer metal plate shell with a composite. Through modeling and simulation, the authors revealed that aperture spacing is the key to enhancing shielding effectiveness, not the thickness of the absorber itself. This provides the insight that offsetting apertures on the inner and outer shielding plates can significantly improve the overall shielding performance, offering an effective design guideline for shielding facilities. Additionally, Park, H.H., et al. (Contribution 9) experimentally analyzed the impact of the air gap between magnetic sheets and the microstrip test board on near-field magnetic shielding effectiveness (NSE). It revealed the dual effects of the air gap on shielding effectiveness and proposed a comprehensive physical explanation for these dual effects. This finding provides critical engineering guidance. Moreover, Park, H.H., et al. (Contribution 10) proposed a method to predict its NSE by measuring the relative permeability of ferrite sheets. Based on the measured NSE and relative permeability data, eight regression models were developed, which significantly improved both prediction efficiency and accuracy.

Li, S., et al. (Contribution 11) investigates damage mechanisms in silicon solar cells under nanosecond pulsed-laser irradiation, establishing dynamic damage evolution models. This study bridges a critical gap in the characterization of laser-induced transient pulse signatures and establishes a novel analytical framework for elucidating damage progression pathways.

Single-Photon Avalanche Diodes (SPADs) have gained prominence in quantum communications and LiDAR systems owing to their single-photon sensitivity, yet radiation-induced degradation poses critical reliability challenges for space applications. Xun, M., et al. (Contribution 12) pioneers a mission-specific approach by replicating the 4-year Geosynchronous Transfer Orbit (GTO) proton environment with tailored shielding configurations. Through a comparative analysis of gamma and proton damage mechanisms under equivalent total doses, this study conducted an in-depth examination by closely correlating macroscopic performance degradation with microscopic physical mechanisms.

Amid rapid deployment of ultra-high-voltage flexible DC transmission infrastructure, growing demand exists for high-altitude applications of flexible DC technology. Radiation tolerance of DC converters (core system components) presents critical engineering challenges. Yang, L., et al. (Contribution 13) pioneers a ground-based neutron radiation test platform simulating high-altitude conditions to quantify failure rate evolution in three critical power devices. By integrating operational radiation data to develop failure prediction models, this study establishes definitive radiation-safe operating boundaries for power devices for the first time, delivering solutions for engineering design and reliability validation.

In conclusion, driven by the accelerating demand for radiation-hardened high-performance electronics in space and nuclear domains, research on radiation effects in advanced devices and integrated circuits has expanded at an unprecedented pace. A comprehensive grasp of the underlying physics and the continuous invention of mitigation strategies are now imperative. The contributions gathered in this Special Issue illustrate the remarkable breadth of current efforts, spanning fundamental nuclear and solid-state physics, system-level modeling, and novel hardening architectures. Looking forward, the third edition of "Radiation Effects of Advanced Electronic Devices and Circuits" will continue to concentrate on next-generation nodes and beyond-CMOS devices such as gate-all-around nanosheets and two-dimensional-material FETs. Attention will also shift to heterogeneous three-dimensional-stacked and chiplet-in-package systems where single-event charge sharing effect and system-level fault propagation dominate. Equally critical is the assurance of AI-enabled space electronics through radiation-aware training and inference frameworks, on-orbit self-healing neuromorphic accelerators, and formal verification of autonomous controllers for safety-critical missions.

Funding: This research was funded by the National Natural Science Foundation of China (Nos. 12205052; 62174180), CAS Talent Program Youth Project (No. E129193YR0), the Science Foundation for Indigenous Innovation of National University of Defense Technology (Grant Number 24-ZZCX-ZXGC-10), and the Guangdong Basic and Applied Basic Research Foundation (No. 2025A1515011606).

Conflicts of Interest: The authors declare no conflicts of interest.

List of Contributions:

1. Cai, Y.; Cai, M.; Wu, Y.; Lu, J.; Bian, Z.; Liu, B.; Cui, S. Evaluation and Mitigation of Weight-Related Single Event Upsets in a Convolutional Neural Network. *Electronics* **2024**, *13*, 1296.
2. Fu, W.; Ma, T.; Lei, Z.; Peng, C.; Zhang, H.; Zhang, Z.; Xiao, T.; et al. Temperature Dependence of Total Ionizing Dose Effects of β-Ga_2O_3 Schottky Barrier Diodes. *Electronics* **2024**, *13*, 2215.
3. Zhao, P.; Tan, X.; Fu, W.; Ma, T. Investigation of Electrical Performance Degradation of β-Ga_2O_3 Schottky Barrier Diodes Induced by X-Ray and Neutron Irradiation. *Electronics* **2025**, *14*, 1343.
4. Huang, H.; Wu, Z.; Peng, C.; Shen, H.; Wu, X.; Yang, J.; Lei, Z.; et al. Comprehensive Study of Proton and Heavy Ion-Induced Damages for Cascode GaN-Based HEMTs. *Electronics* **2025**, *14*, 2653.
5. Zhao, X.; Wang, H.; Zhang, Y.; Chen, Y.; Cheng, S.; Wang, X.; Peng, F.; et al. Model Parameters and Degradation Mechanism Analysis of Indium Phosphide Hetero-Junction Bipolar Transistors Exposed to Proton Irradiation. *Electronics* **2024**, *13*, 1831.
6. He, Y.; Gao, R.; Ma, T.; Zhang, X.; Zhang, X.; Yang, Y. Total Ionizing Dose Effects on Lifetime of NMOSFETs Due to Hot Carrier-Induced Stress. *Electronics* **2025**, *14*, 2563.
7. Zheng, S.; Zhang, Z.; Ye, J.; Lu, X.; Lei, Z.; Liu, Z.; Geng, G.; Zhang, Q.; Zhang, H.; Li, H. Experimental Study of the Impact of Temperature on Atmospheric Neutron-Induced Single Event Upsets in 28 Nm Embedded SRAM of SiP. *Electronics* **2024**, *13*, 2012.
8. Kwon, J.H.; Hyoung, C.-H.; Park, H.H. Analysis of Shielding Performance in Double-Layered Enclosures with Integrated Absorbers. *Electronics* **2024**, *13*, 4345.
9. Park, H.H.; Song, E.; Kim, J.; Kim, C. Impact of Air Gaps Between Microstrip Line and Magnetic Sheet on Near-Field Magnetic Shielding. *Electronics* **2024**, *13*, 4313.
10. Park, H.H.; Lee, H.; Hwang, D.-K. Regression Analysis for Predicting the Magnetic Field Shielding Effectiveness of Ferrite Sheets. *Electronics* **2025**, *14*, 310.
11. Li, S.; Huang, L.; Ye, J.; Hong, Y.; Wang, Y.; Gao, H.; Cui, Q. Study on Radiation Damage of Silicon Solar Cell Electrical Parameters by Nanosecond Pulse Laser. *Electronics* **2024**, *13*, 1795.
12. Xun, M.; Li, Y.; Liu, M. Comparison of Proton and Gamma Irradiation on Single-Photon Avalanche Diodes. *Electronics* **2024**, *13*, 1086.
13. Yang, L.; Zhang, Z.; Zhou, Y.; Wang, D.; Peng, C.; Zhang, H.; Lei, Z.; Zhang, Z.; Fu, W.; Ma, T. Effect of Cosmic Rays on the Failure Rate of Flexible Direct Current Converter Valves in High-Altitude Environment. *Electronics* **2024**, *13*, 4790.

References

1. Wang, H.; Dai, X.; Ibrahim, Y.M.Y.; Sun, H.; Nofal, I.; Cai, L.; Guo, G.; Shen, Z.; Chen, L. A layout-based rad-hard DICE flip-flop design. *J. Electron. Test.* **2019**, *35*, 111–117. [CrossRef]
2. Artola, L.; Gaillardin, M.; Hubert, G.; Raine, M.; Paillet, P. Modeling single event transients in advanced devices and ICs. *IEEE Trans. Nucl. Sci.* **2015**, *62*, 1528–1539. [CrossRef]
3. Reed, R.A.; Weller, R.A.; Mendenhall, M.H.; Fleetwood, D.M.; Warren, K.M.; Sierawski, B.D.; King, M.P.; Schrimpf, R.D.; Auden, E.C. Physical processes and applications of the Monte Carlo radiative energy deposition (MRED) code. *IEEE Trans. Nucl. Sci.* **2015**, *62*, 1441–1461. [CrossRef]
4. Fleetwood, D.M. Radiation effects in a post-Moore world. *IEEE Trans. Nucl. Sci.* **2021**, *68*, 509–545. [CrossRef]
5. Fleetwood, Z.E.; Lourenco, N.E.; Ildefonso, A.; Warner, J.H.; Wachter, M.T.; Hales, J.M.; Tzintzarov, G.N.; Roche, N.J.-H.; Khachatrian, A.; Buchner, S.P.; et al. Using TCAD modeling to compare heavy-ion and laser-induced single event transients in SiGe HBTs. *IEEE Trans. Nucl. Sci.* **2017**, *64*, 398–405. [CrossRef]
6. Chatterjee, I.; Narasimham, B.; Mahatme, N.N.; Bhuva, B.L.; Reed, R.A.; Schrimpf, R.D.; Wang, J.K.; Vedula, N.; Bartz, B.; Monzel, C. Impact of technology scaling on SRAM soft error rates. *IEEE Trans. Nucl. Sci.* **2014**, *61*, 3512–3518. [CrossRef]
7. Cai, C.; Hu, M.; Shen, L.; Yu, J.; Chen, G. Design and evaluation of a radiation-hardened FDSOI SRAM with high-reliable elements and power management circuits for space application. *IEEE Trans. Aerosp. Electron. Syst.* **2025**, *61*, 5057–5066. [CrossRef]

8. Cai, C.; Liu, Y.; Hu, M.; Chen, G.; Yu, J. Characterization of LDO induced increment of SEE sensitivity for 22-nm FDSOI SRAM. *IEEE Trans. Device Mater. Rel.* **2023**, *23*, 537–543. [CrossRef]
9. Caron, P.; Inguimbert, C.; Artola, L.; Bezerra, F.; Ecoffet, R. New SEU modeling method for calibrating target system to multiple radiation particles. *IEEE Trans. Nucl. Sci.* **2020**, *67*, 44–49. [CrossRef]
10. De Sio, C.; Azimi, S.; Portaluri, A.; Sterpone, L. SEU evaluation of hardened-by-replication software in RISC-V soft processor. In Proceedings of the 2021 IEEE International Symposium on Defect and Fault Tolerance in VLSI and Nanotechnology Systems (DFT), Athens, Greece, 6–8 October 2021; pp. 1–6.
11. Hoque, K.A.; Mohamed, O.A.; Savaria, Y. Dependability modeling and optimization of triple modular redundancy partitioning for SRAM-based FPGAs. *Reliab. Eng. Syst. Saf.* **2019**, *182*, 107–119. [CrossRef]
12. Kobayashi, D.; Hirose, K.; Sakamoto, K.; Tsuchiya, Y.; Okamoto, S.; Baba, S.; Shindou, H.; Kawasaki, O.; Makino, T.; Ohshima, T. An SRAM SEU cross section curve physics model. *IEEE Trans. Nucl. Sci.* **2022**, *69*, 232–240. [CrossRef]
13. Mousavi, M.; Pourshaghaghi, H.R.; Tahghighi, M.; Jordans, R.; Corporaal, H. A generic methodology to compute design sensitivity to SEU in SRAM-based FPGA. In Proceedings of the 21st Euromicro Conference on Digital System Design (DSD), Prague, Czech Republic, 29–31 August 2018; pp. 221–228.
14. Villalta, I.; Bidarte, U.; Gómez-Cornejo, J.; Jiménez, J.; Lázaro, J. SEU emulation in industrial SoCs combining microprocessor and FPGA. *Reliab. Eng. Syst. Saf.* **2018**, *170*, 53–63. [CrossRef]
15. Wang, J.; Prinzie, J.; Coronetti, A.; Thys, S.; Alia, R.G.; Leroux, P. Study of SEU sensitivity of SRAM-based radiation monitors in 65-nm CMOS. *IEEE Trans. Nucl. Sci.* **2021**, *68*, 913–920. [CrossRef]
16. Xiong, Y.; Pieper, N.J.; Feeley, A.T.; Narasimham, B.; Ball, D.R.; Bhuva, B.L. Single-event upset cross-section trends for D-FFs at the 5- and 7-nm bulk FinFET technology nodes. *IEEE Trans. Nucl. Sci.* **2023**, *70*, 381–386. [CrossRef]

Disclaimer/Publisher's Note: The statements, opinions and data contained in all publications are solely those of the individual author(s) and contributor(s) and not of MDPI and/or the editor(s). MDPI and/or the editor(s) disclaim responsibility for any injury to people or property resulting from any ideas, methods, instructions or products referred to in the content.

electronics

MDPI

Article

Comparison of Proton and Gamma Irradiation on Single-Photon Avalanche Diodes

Mingzhu Xun [1,2,3], Yudong Li [1,2,*] and Mingyu Liu [1,2,3]

1 State Key Laboratory of Functional Materials and Devices for Special Environmental Conditions, Xinjiang Technical Institute of Physics and Chemistry, Chinese Academy of Sciences, Urumqi 830011, China; xunmz@ms.xjb.ac.cn (M.X.); liumingyu21@mails.ucas.ac.cn (M.L.)
2 Xinjiang Key Laboratory of Extreme Environment Electronics, Urumqi 830011, China
3 University of Chinese Academy of Sciences, Beijing 100049, China
* Correspondence: lydong@ms.xjb.ac.cn

Abstract: In this paper, the effects of proton and gamma irradiation on reach-through single-photon avalanche diodes (SPADs) are investigated. The I–V characteristics, gain and spectral response of SPAD devices under proton and gamma irradiation were measured at different proton energies and irradiation bias conditions. Comparison experiments of proton and gamma irradiation were performed in the radiation environment of geosynchronous transfer orbit (GTO) with two different radiation shielding designs at the same total ionizing dose (TID). The results show that after 30 MeV and 60 MeV proton irradiation, the leakage current and gain increase, while the spectral response decreases slightly. The leakage current degradation is more severe under the "ON"-bias condition compared to the "OFF"-bias condition, and it is more sensitive to the displacement radiation damage caused by protons compared to gamma rays under the same TID. Further analysis reveals that the non-elastic and elastic cross-section of protons in silicon is 1.05×10^5 times greater than that of gamma rays. This results in SPAD devices being more sensitive to displacement radiation damage than ionizing radiation damage. Under the designed shielding conditions, the leakage current, gain and spectral response parameters of SPADs do not show significant performance degradation in the orbit.

Keywords: SPADs; proton irradiation; gamma irradiation; I–V characteristics

1. Introduction

A SPAD is a photodetector capable of detecting and counting single photons. By reversing the bias of the PN junction and operating above the breakdown voltage, photoelectrons generated by the photoelectric effect experience continuous multiplication in a strong electric field when a single photon impacts the multiplication region of the SPAD, resulting in an amplified photocurrent [1,2]. The photocurrent can be converted into a voltage signal for preamplifying or counting by an active or passive external quenching circuit. A SPAD has the advantages of high sensitivity, high gain, high temporal resolution, magnetic resistance and low noise for single-photon detection, and in recent years, it has been widely used in a variety of weak optical signal detection fields, including star-ground quantum communication, quantum key distribution, quantum computation, and high-energy particle detection in space, atmospheric pollution monitoring, astronomical observation, photoelectric imaging and other space missions [3–5].

In 1963, R.H. Haitz et al. clarified the working principle of SPADs, explained the physical phenomena generated by photon pulses, dark current pulses and residual pulses, and proposed the theoretical model of photon pulses [6,7]. They also designed a SPAD device model. In 1999, McIntyre et al. established a comprehensive theoretical model for the impact ionization process of SPADs [8,9]. After 60 years of rapid development, the materials, structures and circuits of SPADs have been optimized, and the type and performance of

SPADs have been significantly improved, accelerating the large-scale application of single-photon detectors in fields such as quantum communication, light detection and ranging (LIDAR), and medical detection [5,10–13]. The key to the performance improvement is the availability of high-quality materials with higher purity and fewer defects, as well as novel SPAD structures. Structurally, the types of detectors that have been developed include Si-SPADs, InGaAs detectors, Ge detectors, SiC detectors, superconducting nanowire single-photonic detectors (SNSPDs) and quantum-dot optically gated field-effect transistors (QDOGFETs) [14–19]. Different SPAD materials have different sensitivities to wavelengths. For example, GaN-based SPADs are sensitive to ultraviolet (UV) light and are typically used to measure weak UV light. Due to the bandwidth of silicon, Si-SPADs have a limited operating range and can only be used in the visible and near-infrared wavelengths. For infrared wavelengths above 1100 nm, materials with a bandgap of less than 1.1 eV should be used, typically In-GaAs and InP materials [20]. Structural innovations include back-illuminated SPADs, vertical double-junction SPADs, substrate-isolated SPADs, 3D-stacked SPADs, non-contact ring guard SPADs and high-detection-efficiency SPADs [21,22]. These novel materials and structures further improve SPADs' detection efficiency and reduce dark counts.

With the large-scale commercialization and application of low-cost devices in the space field, SPADs are inevitably exposed to proton and electron radiation in the space radiation environment, leading to radiation damage to the SPAD. Recent studies have shown that the radiation effect on Si-based devices mainly causes TID and displacement damage dose (DDD) [23,24]. The energy loss of incident particle ionization leads to the generation of electron hole pairs in irradiated materials, while non-ionization energy loss leads to phonon generation and lattice atomic displacement, resulting in ionization and displacement damage [25,26]. These damages can form movable charged defects, interface states defects and Frenkel defects in silicon-based devices, which can act as carrier traps and recombination centers [27,28]. This damage can increase the leakage current, dark count rate (DCR) and equivalent noise of the SPAD, and potentially affect the performance and lifetime of the SPAD.

To investigate the radiation effects on SPADs, we conducted a detailed analysis of the radiation damage inflicted by protons and gamma rays on SPAD devices. The radiation environments of specific orbit altitude and radiation shielding designs were considered. Furthermore, the differences and similarities in the degradation of SPAD parameters were compared. In this experiment, back-illuminated SPADs were selected. The changes in dark current, gain and spectral response of the SPAD after proton and gamma irradiation were analyzed in detail. The degradation of the SPAD parameters under equivalent TID for proton and gamma radiation were compared. From the perspectives of the number of displacement damage defects and collision cross-sections, the SPAD degradation mechanisms of proton and gamma radiation were explained.

2. Experimental Design

The cross-section of the back-illuminated SPADs used in this experiment is shown in Figure 1 [29]. The SPADs have a thick depletion layer structure, and the multiplication region is composed of a PN junction formed by the contact between phosphorus-doped n+-type and boron-doped p-type semiconductors, as well as a lightly doped phosphorus-doped n-type guard ring around the edges of the PN junction [30]. The guard ring structure prevents edge breakdown, reduces surface tunneling, and improves photon detection efficiency. The distance from the wafer surface to the depletion region of the PN junction is reduced to 30–40 μm in thickness through a wafer back etching process, and a low-doped, low-resistance light absorption region is formed using a boron diffusion process. When photons impact in the absorption layer and generate photoelectrons, the photoelectrons drift into the multiplication region under the reverse electric field. The strong electric field ($E > 1 \times 10^5$ V/cm) in the multiplication region accelerates the photoelectrons to obtain sufficient kinetic energy, and the photoelectrons transfer energy to the electrons of new lattice atoms, ionizing them into secondary electrons through the Coulomb interaction. The photoelectrons and secondary electrons continuously multiply in the multiplication region, forming a sustainable multiplication current. The multiplication current can be converted into a voltage signal for subsequent signal amplification and pulse counting by an external active or passive quenching circuit. The light-absorbing region of this SPAD structure is relatively thick, usually 30–40 μm. A high breakdown voltage is required to achieve the avalanche breakdown condition, usually above 200 V [31]. Due to the use of epitaxial manufacturing processes, the lattice defects, point defects and impurity concentrations in the SAPD are lower. But the remaining defects in the multiplication region cause Shockley–Read–Hall (SRH) recombination and trap-assisted tunneling, resulting in the random release of carriers and an increase in the DCR [29]. Therefore, the SPAD needs to operate at a low temperature of −20 °C to reduce the DCR. The breakdown voltage of the SPAD selected for this experiment is about 225 V, and the reverse leakage current is about 7 nA. A reach-through structural SPAD was selected for gamma and proton radiation effect research due to its simple structure, mature manufacturing process and complete theoretical model for analyzing I–V and spectrum characteristics [31,32].

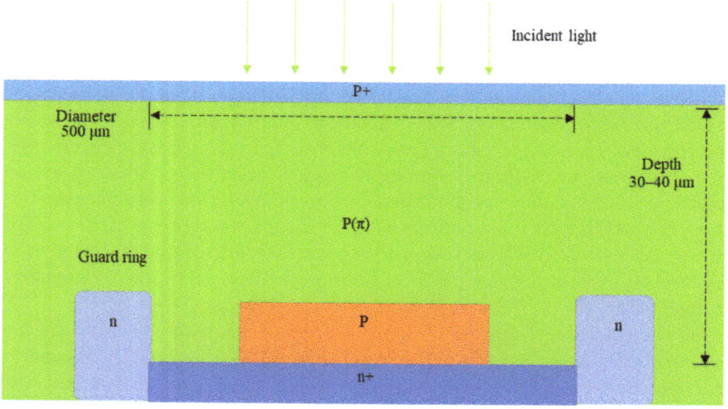

Figure 1. Cross-section of the reach-through structure of the SPAD.

In order to analyze the effect of TID and DDD on the SPAD in a spatial radiation environment, this experiment was designed based on the radiation environment at the GTO, with an altitude of 35,786 km, corresponding to a mission duration of four years in the orbit. Two shielding materials were designed: a 3.705 mm (1.0 g/cm^2) aluminum shield material, and 8.234 mm Al with a 0.467 mm Ta shield material (3.0 g/cm^2 Ta to Al with a mass ratio of 35%). After the simulation of space radiation particle transport in the shielding material by Geant4, which is a toolkit for the simulation of particles through matter invented by the European Organization for Nuclear Research (CERN) [33–35], it was calculated that the energies corresponding to the differential energy spectrum peaks of the protons (including primary and secondary protons) generated by solar protons after passing through the two shielding materials are approximately 30 MeV and 60 MeV, as shown in Figure 2. The TID and DDD were also calculated as shown in Table 1. In addition to particles produced by solar protons, protons and electrons from the Van Allen radiation belts and solar flare outbursts also contribute to TID and DDD. The contribution of radiation belts and solar flare outbursts to TID and DDD was taken into account in the design of this experiment. Therefore, the TID and DDD corresponding to the total proton flux used in the experiment exceed the shielding design values shown in Table 1, as detailed in Table 2.

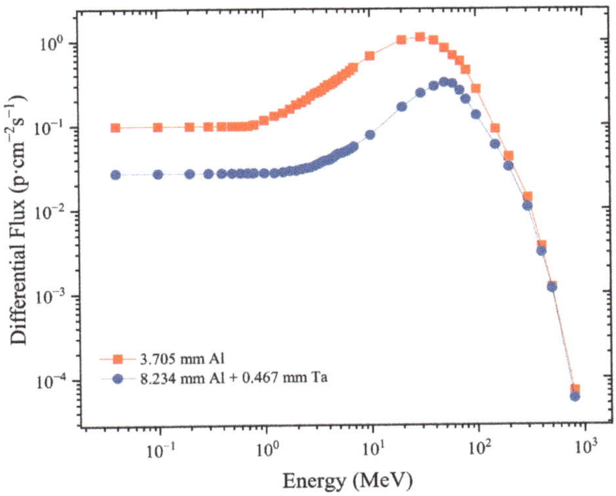

Figure 2. Proton differential flux of two designed shielding materials.

Table 1. TID and DDD of two designed shielding materials at the GTO for 4 years.

Designed Shielding Material	TID (krad)	DDD (MeV/g)
3.705 mm Al	1.35	2.84×10^7
8.234 mm Al + 0.467 mm Ta	0.346	7.02×10^6

Table 2. Proton fluence, LET, NIEL, TID and DDD of 30 MeV and 60 MeV protons used in experiment.

Proton Energy (MeV)	LET (MeV/(g/cm^2))	NIEL (MeV/(g/cm^2))	Fluence (p/cm^2)	TID (krad)	DDD (MeV/g)
30	14.76	5.63×10^{-3}	1.00×10^9	0.235	5.63×10^6
			5.00×10^9	1.175	2.82×10^7
			1.00×10^{10}	2.35	5.63×10^7
			5.00×10^{10}	11.75	2.82×10^8
			1.00×10^{11}	23.5	5.63×10^8
60	8.6	2.89×10^{-3}	1.00×10^9	0.138	2.89×10^6
			5.00×10^9	0.69	1.45×10^7
			1.00×10^{10}	1.38	2.89×10^7
			5.00×10^{10}	6.9	1.45×10^8
			1.00×10^{11}	13.8	2.89×10^8

The proton testing was conducted at the Proton Radiation Effect simulation test Facility (PREF) of the Xinjiang Institute of Physics and Chemistry, Chinese Academy of Sciences. The proton energy was 30 MeV and 60 MeV and the beam area was 2 cm × 2 cm, with a proton fluence of 4.79×10^8 p/cm^2. The proton uniformity was within ±5%. The proton fluence, line energy transport (LET), non-ionizing energy loss (NIEL), TID and DDD of the experiment are shown in Table 2. The LET data were obtained from the National Institute of Standards and Technology (NIST)'s stopping power and range tables for protons (PSTAR) program, and the NIEL data were obtained from reference [36]. The gamma irradiation was conducted at the Co-60 gamma source of the Xinjiang Institute of Physics and Chemistry, Chinese Academy of Sciences. The dose rate was 50 rad(Si)/s, with a cumulative dose of 70 krad(Si).

Three bias conditions were applied during the irradiation: "$ON_{M=100}$"-bias condition, "$ON_{M=10}$"-bias condition and "OFF"-bias condition. The "$ON_{M=100}$" bias applies an operating voltage of 215 V, and at this voltage, the gain of the SPAD is 100 (M = 100). The "$ON_{M=10}$" bias applies an operating voltage of 180 V and the gain of the SPAD is 10 (M = 10). The "OFF" bias applies an operating voltage of 0 V and all pins are shorted together and grounded. All parameter measurements were completed within 2 h after irradiation.

The schematic diagram of the electrical parameter testing system used is shown in Figure 3. The halogen lamps provide a stable light source with a wavelength of 350–1100 nm. After passing through the filter, a single wavelength light is generated in the spectrograph. The integrating sphere performs multiple diffuse reflections on the incident light to improve the uniformity of the light incident on the test sample, ensuring photostability during sample testing. In terms of parameter measurements, the Keysight semiconductor parameter analyzer was used to measure the I–V characteristics of the device before and after irradiation, and an oscilloscope was used to measure and observe the SPAD's pulse output waveform. The external test circuit consisted of a passive quenching circuit with a 50 kΩ resistor. An external counter was used to accurately measure the dark count.

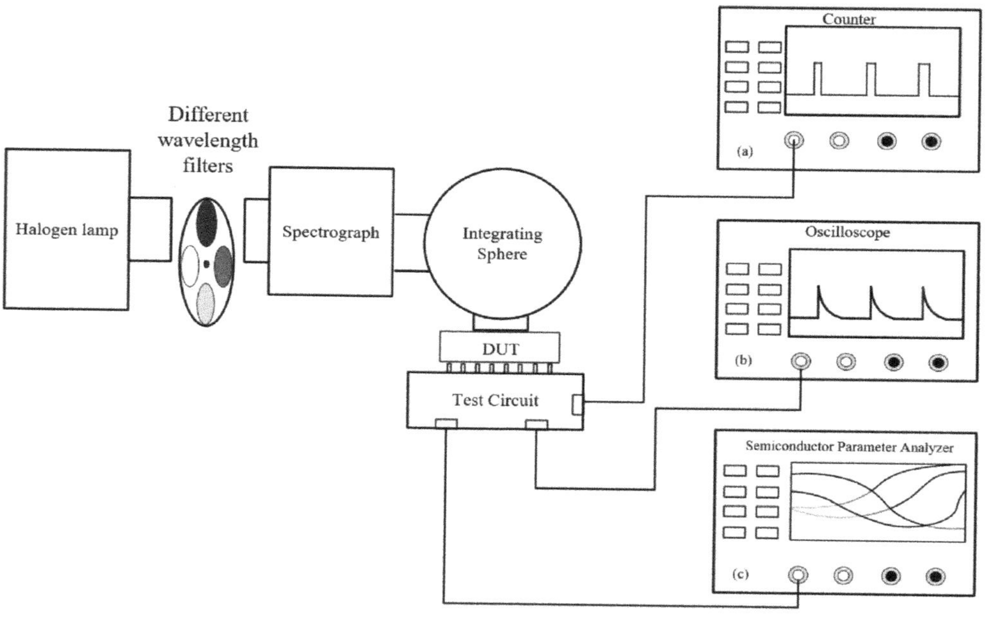

Figure 3. The SPAD parameter testing system with (**a**) a counter for the dark count rate (DCR) measurement, (**b**) an oscilloscope for pause shape observations and (**c**) a semiconductor parameter analyzer for I–V characteristic measurements [37].

3. Results

3.1. I–V Characteristics

The I–V characteristics of the SPAD before and after gamma radiation are shown in Figure 4. Figure 4a shows that the leakage current increases with increasing radiation dose under the "OFF"-bias condition. This is measured in a dark environment. At a reverse voltage of 200 V, the leakage current increases from 5.79 nA before irradiation to 6.8 nA after accumulating 70 krad(Si) TID, an increase of 1.01 nA. When the accumulated TID reaches 2.35 krad(Si), which corresponds to a proton fluence of 1×10^{10} p/cm^2, the leakage current at a reverse voltage of 200 V increases to 6.173 nA, an increase of 0.383 nA. This indicates that the leakage current does not increase significantly under "OFF"-bias conditions. However, under the "$ON_{M=100}$"-bias condition, as shown in Figure 4b, the leakage current increases from 6.39 nA to 20.94 nA after accumulating a TID of 30 krad(Si) at a reverse voltage of 200 V. When the accumulated TID reaches 70 krad(Si), the leakage current increases to 12.26 μA. This implies that the SPAD is severely damaged. The increase in leakage current of the "$ON_{M=100}$"-bias irradiated SPAD is more significant compared to the "OFF"-bias condition at the same TID.

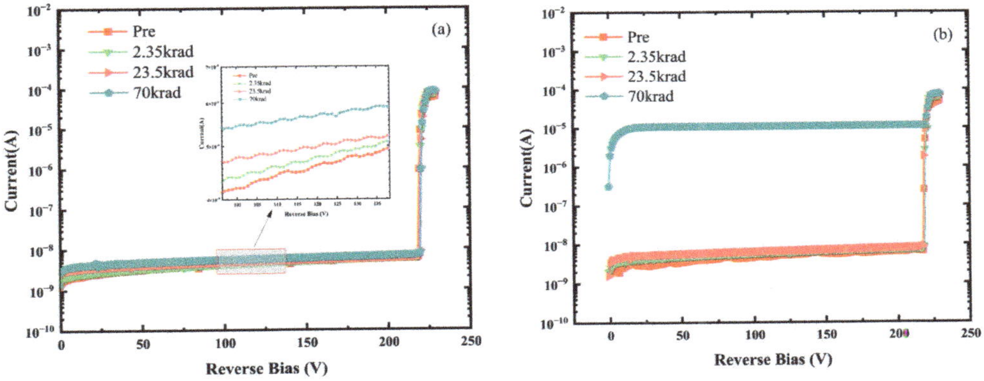

Figure 4. I–V characteristics of SPAD after γ irradiation under the (**a**) "OFF"-bias condition and (**b**) "$ON_{M=100}$"-bias condition.

The I–V characteristics of the SPAD before and after 30 MeV proton irradiation are shown in Figure 5. As shown in Figure 5a, the leakage current increases slightly with the increase in proton irradiation fluence under the "OFF"-bias condition. At a reverse voltage of 80 V, the leakage current increases from 3.49 nA to 4.66 nA at a fluence of 5×10^{10} p/cm^2. However, when the reverse voltage reaches 117 V, the leakage current exhibits a "step" increase, similar to the "punch-through" phenomenon between the multiplication region and the absorption region [38]. At a reverse voltage of 200 V, the reverse leakage current increases from 5.42 nA before irradiation to 24.7 nA after irradiation. It can be seen that as the reverse voltage increases, the leakage current also increases, but the breakdown voltage of the SAPD device does not change significantly before and after irradiation. The results of proton irradiation under the "$ON_{M=100}$"-bias condition are shown in Figure 5b. The leakage current also gradually increases with the accumulation of fluence. However, for the 1×10^{10} p/cm^2 proton fluence required by the on-orbit shielding design, the leakage current of the SPAD device increases from 5.68 nA before irradiation to 14.3 nA after irradiation at an operating voltage of 200 V. But when the proton fluence accumulates to 5×10^{10} p/cm^2, the leakage current suddenly increases to 1.82 μA, which means there is serious damage. This indicates that under proton irradiation, with an "$ON_{M=100}$"-bias condition, the damage to the SPAD is more severe compared to that caused by the "OFF"-bias condition.

To further analyze the effect of bias conditions on the reverse leakage current under different proton irradiation energies, we conducted proton irradiation with 30 MeV and 60 MeV protons at a cumulative fluence of 1×10^{11} p/cm^2. Three bias conditions were applied during the irradiation: the "$ON_{M=100}$"-bias condition, "$ON_{M=100}$"-bias condition and "OFF"-bias condition. The measured I–V characteristics are shown in Figure 6. From the test results, it can be observed that under the same proton energy, the leakage current increases with the increase in bias gain, indicating that the number of defects produced by proton incidence in the SPAD increase with the increase in the electric field in the multiplication region. Under the "OFF"-bias condition, both proton energies exhibit a "step" increase near the voltage of 117 V, which does not occur under the conditions of gain of 10 and 100. The reason may be that under the condition of no electric field, the charged defects generated during irradiation cannot migrate due to the absence of an electric field, resulting in local accumulation. This creates a local built-in electric field opposite to the direction of the electric field during device operation. This built-in electric field hinders the transport of charge carriers, reducing the leakage current. As the voltage increases to a certain level, the charges trapped by the defects are released under the strong electric field, reducing or even eliminating the built-in electric field, leading to a sudden increase in the leakage current. Under both 30 MeV and 60 MeV irradiation energies, the reverse leakage current under the three bias conditions was higher for 30 MeV proton irradiation than for 60 MeV proton irradiation. This is due to the higher LET and NIEL values for 30 MeV proton irradiation compared to 60 MeV, as shown in Table 2. The higher defect production by 30 MeV protons results in more severe radiation damage to the SPAD. Further analysis reveals that under the condition of gain of 100, the effect of proton displacement damage dose (DDD) on the leakage current of the SPAD is greater than the effect of total ionizing dose (TID). The TID for 30 MeV protons at a fluence of 1×10^{11} p/cm^2 is 23.5 krad(Si). However, as shown in Figure 4b, the leakage current after gamma irradiation at this dose is less than 10 nA. In contrast, when proton irradiation is carried out at a fluence of 1×10^{11} p/cm^2, which corresponds to a TID of 23.5 krad(Si) and a DDD of 5.63×10^8 MeV/g, the leakage current increases by a factor of 1000 to more than 10 μA. This demonstrates that compared to total ionizing damage, the leakage current of the SPAD is more sensitive to displacement damage.

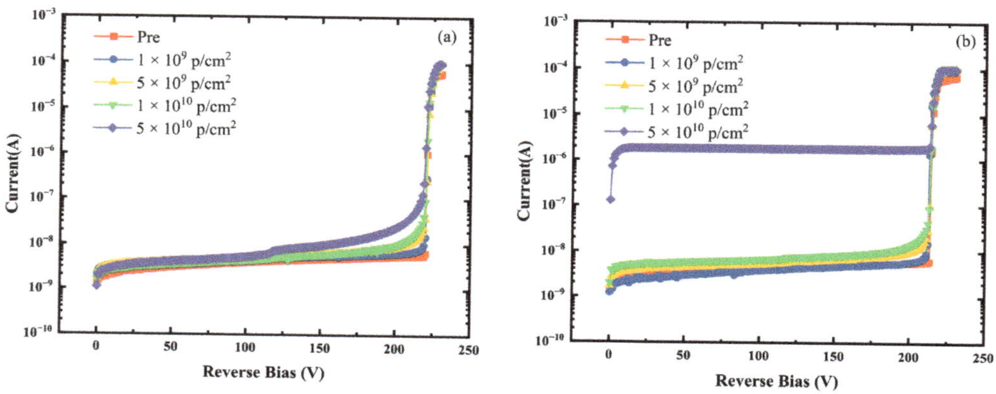

Figure 5. I–V characteristics of SPAD after 30 MeV proton irradiation under the (**a**) "OFF"-bias condition and (**b**) "$ON_{M=100}$"-bias condition.

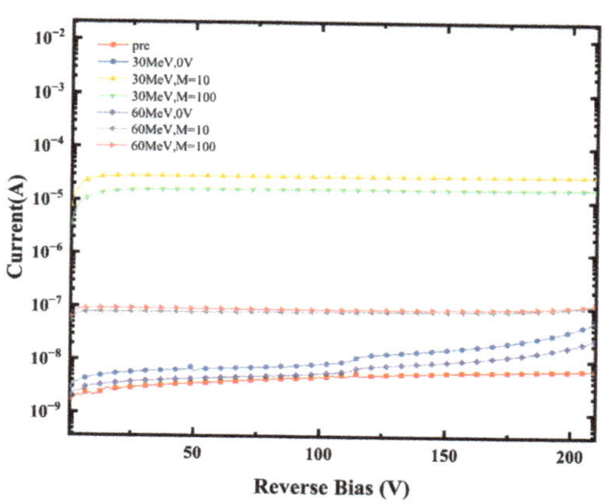

Figure 6. I–V characteristics of SPAD after proton irradiation under different bias conditions of 30 MeV and 60 MeV.

3.2. Multiplication

The variation in gain versus reverse voltage after gamma and proton irradiation is shown in Figure 7. Here, the gain is defined as the ratio of the photocurrent with bias voltage to the photocurrent without bias voltage. Figure 7a represents the gain curve after gamma irradiation. It can be observed that under the "OFF"-bias condition, as the accumulated TID increases, the SPAD gain remains almost unchanged. This indicates that at gamma irradiation doses up to 70 krad(Si), the photocurrent generation capability is not significantly improved compared with that before irradiation. The defects produced by gamma irradiation do not affect the carrier drift process in the absorption region or the avalanche process in the multiplication region of the SPAD device. Therefore, the gain remains almost unchanged after gamma irradiation.

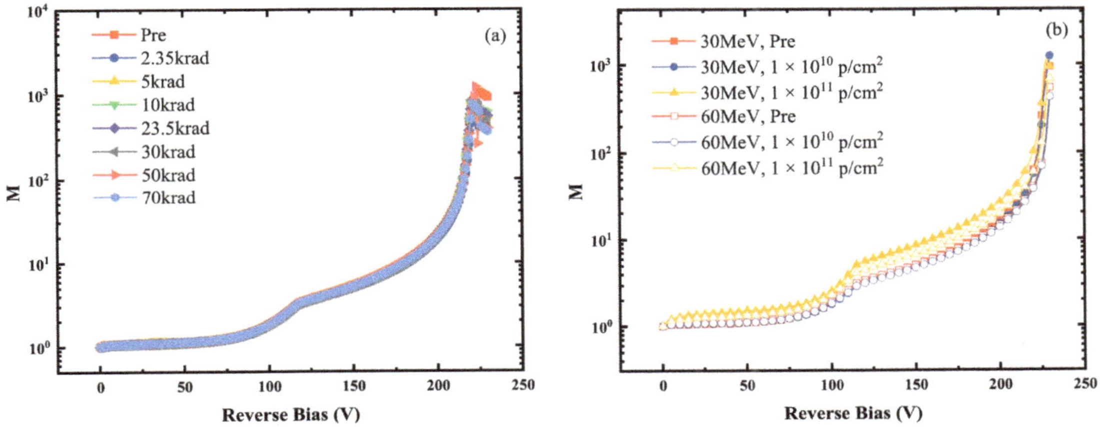

Figure 7. Multiplication of SPAD after (**a**) γ irradiation and (**b**) 30 MeV and 60 MeV proton irradiation.

Figure 7b shows the gain variation at different reverse voltages after 30 MeV and 60 MeV proton irradiation. From the figure, it can be seen that at a fluence of 1×10^{10} p/cm^2, the gain of the SPAD remains almost constant at both energies. However, when the accumulated fluence reaches 1×10^{11} p/cm^2, the gain starts to increase, and the gain after 30 MeV proton irradiation is more significant than that after 60 MeV proton irradiation, which indicates that 30 MeV protons cause more serious radiation damage to the SPAD. By observing the change in photocurrent and dark current before and after irradiation, it was found that both currents increased after irradiation. However, the dark current I_{dark} and the $I_{unit,dark}$ both increased significantly compared with those before irradiation. The increase in the $I_{unit,dark}$ is more significant, resulting in a decrease in the photocurrents without bias voltage after irradiation, and a consequent increase in the gain. The gain (M) is calculated as follows:

$$M = \frac{I_{photo} - I_{dark}}{I_{unit,photo} - I_{unit,dark}}$$

In the formula, I_{photo} is the photocurrent measured in the presence of light and with bias voltage, I_{dark} is the leakage current measured in the absence of light but with bias voltage, $I_{unit,photo}$ is the photocurrent measured in the presence of light but without bias voltage, and $I_{unit,dark}$ is the leakage current measured in the absence of light and without bias voltage.

3.3. Spectral Response

Figure 8a displays the spectral response after gamma irradiation under the "ON$_{M=100}$"-bias condition. It is worth noting that the spectral response of the SPAD with an incident wavelength of less than 700 nm remains relatively stable during irradiation with a total cumulative dose of 30 krad(Si). However, in the wavelength range of 700–900 nm, the spectral response fluctuates, showing an increase under a TID of 30 krad(Si) followed by a decrease under a TID of 50 krad(Si). This behavior may be attributed to variations in carrier release and multiplication caused by mobile charged defects generated by ionizing radiation, inherent device defects and displacement damage defects resulting from a high TID.

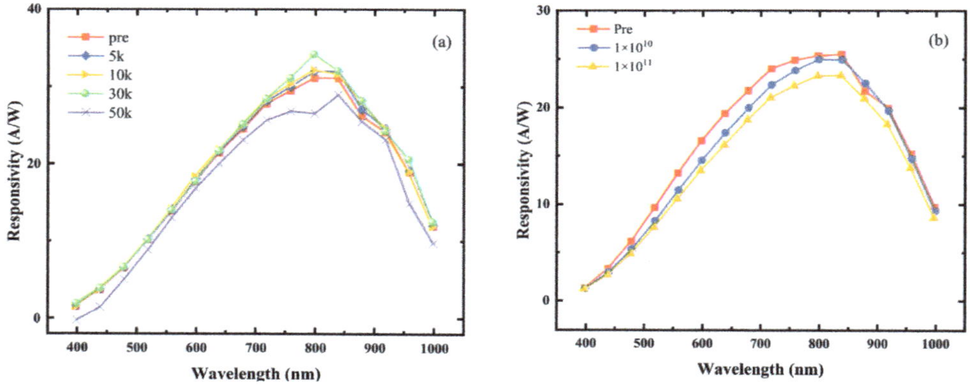

Figure 8. Spectral response of SPAD after (**a**) γ irradiation and (**b**) 60 MeV proton irradiation under "OFF"-bias condition.

Figure 8b shows the spectral response after 60 MeV proton irradiation under the "OFF"-bias condition. When the proton irradiation fluence reaches 1×10^{10} p/cm^2, the spectral response curve decreases significantly at wavelengths of 400–850 nm, and when the proton fluence reaches 1×10^{11} p/cm^2, the spectral response decreases at wavelengths in the range of 400–1000 nm. The photocurrent of the device is reduced due to the effect of displacement damage defects formed by proton irradiation in the photoelectron multiplication process, leading to a decrease in spectral response. Reference [39] reported spectral response degradation after neutron irradiation of avalanche photodiodes (APDs), and the neutron energy was 1 MeV, the neutron flux was 2×10^{13} n/cm^2 and the cumulative DDD was 4.07×10^{10} MeV/g. The peak spectral response at a wavelength of 600 nm decreased from 23.5 A/W to 9.5 A/W after irradiation, with a gain of 40. This indicates that displacement damage produced by protons and neutrons would also cause spectral response degradation.

4. Discussion

Based on the previous experimental results, it can be observed that compared to the ionization radiation damage caused by gamma irradiation, the displacement radiation damage inflicted by protons on SPADs leads to more pronounced radiation-induced degradation in the leakage current, gain and spectral response. Previous research has shown that gamma radiation in silicon-based devices primarily produces ionization radiation damage through Compton scattering and displacement damage due to collisions between secondary high-energy electrons and silicon [40]. On the other hand, protons cause ionization radiation damage through ionization and excitation processes, and displacement damage through elastic and inelastic collisions (nuclear reactions) between protons and silicon atoms. These ionization and displacement radiation damage processes generate defects such as vacancy (V_{Si}), substitutional phosphorus (P_{Si}), interstitial oxygen (O_i), double vacancy (V_{Si}-V_{Si}), A-center (V_{Si}-O_i) and E-center (P_{Si}-V_{Si}) [27,28,41–43]. Displacement radiation damage-induced vacancy defects can form deep-level defects within the SPAD's multiplication region, capturing and releasing carriers as generation–

recombination centers. They randomly emit electrons to the multiplication area, causing increased leakage currents.

Due to the more severe parameter degradation observed in SPADs after proton irradiation, the Stopping and Range of Ions in Matter Software 2013 (SRIM-2013) is used to analyze the displacement radiation damage caused by 30 MeV and 60 MeV protons [44]. This software performs Monte Carlo simulations of ion transportation; through detailed calculations, the number of vacancy defects generated per particle, the energy deposition, the particle range and other relevant information can be obtained. The simulation results for 30 MeV and 60 MeV protons are presented in Table 3. It can be seen that 30 MeV protons produce a displacement damage vacancy fluence of 24 n/(ion·mm) in silicon, while 60 MeV protons produce only 10.4 n/(ion·mm). This explains why the parameter degradation after 30 MeV proton irradiation is more severe than that caused by 60 MeV proton irradiation. Additionally, Figure 9 shows the depth dose distribution curves, indicating that although 60 MeV protons have a deeper penetration depth, within the effective thickness of the device (~200 μm), 30 MeV protons exhibit a higher stopping power at the SAPD surface, depositing more energy and generating a higher number of vacancy defects. This leads to more severe parameter degradation under the same proton irradiation conditions.

Table 3. Comparison of the SPAD breakdown voltage.

Proton Energy (MeV)	Total Vacancies (n/(ion·mm))	Total Stopping Power (MeV/g)	Projected Range (mm)
30	24	3.441×10	4.91
60	10.4	2.004×10	16.85

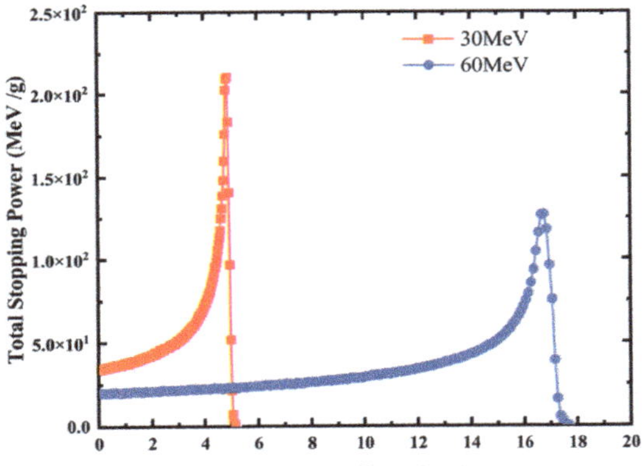

Figure 9. Depth dose of 30 MeV and 60 MeV protons in silicon.

The internal structure of the SPAD devices discussed does not have a large Si-SiO_2 interface. Therefore, migration of ionizing radiation damage defects to the interface due to gamma irradiation does not result in the formation of a large number of interfacial states [45,46]. The parameter changes observed after gamma irradiation are mainly attributed to displacement damage defects formed by energetic secondary electrons generated by the interaction of gamma rays with lattice atoms [25,47].

In fundamental studies of ray–matter interactions, the reaction cross-section is used to characterize the probability of interactions between rays and matter. Physically, it represents the probability of a nuclear reaction occurring when a single particle impinges on a unit area containing only one target nucleus. The unit of the cross-section is $1b = 10^{-24}$ cm^2. By referring to the TENLDL-2023 basic nuclear database published by the Paul Scherrer Institute (PSI) and comparing gamma ray and proton interactions and nuclear reaction processes on silicon materials, we obtain the data presented in Table 4. It can be observed that gamma rays primarily suffer from elastic collisions with silicon atomic nuclei when entering silicon, without non-elastic scattering. This means that the displacement damage is caused by elastic collisions between gamma photons and atomic nuclei, without involving nuclear reactions. The Compton effect between photons and outer shell electrons is not included in this elastic scattering cross-section.

Table 4. Elastic and non-elastic cross-section of gamma rays and protons in silicon.

Incident Particle	Particle Energy (MeV)	Non-Elastic Cross-Section(mb)	Elastic Cross-Section (mb)
gamma	1	0	2.801×10^{-3}
	1.25 (^{60}Co)	0	5.522×10^{-3}
	2	0	1.368×10^{-2}
proton	30	7.428×10^{2}	4.835×10^{-3}
	60	5.778×10^{2}	3.072×10^{-9}

On the other hand, when 30 MeV and 60 MeV protons impact the silicon material, inelastic processes dominate, and the elastic collision cross-section becomes very small as the energy of the incident proton increases. Both elastic and inelastic collisions of incident protons lead to the displacement of silicon atoms and the formation of displacement damage defects. The data in Table 4 show that the scattering cross-section of gamma rays is 5.522×10^{-3} mb, while the elastic and non-elastic scattering cross-section of 60 MeV protons is 5.778×10^{2} mb, which is 1.05×10^{5} times larger than the scattering cross-section of gamma rays.

The interaction process between protons and gamma rays with the silicon nucleus is described above. Secondary electron production by the Compton effect of gamma rays in silicon, as well as the ionization and excitation process of electrons in the atomic shell caused by protons in the presence of Coulomb gravity, are not considered. This is because displacement radiation damage is primarily caused by atoms leaving their lattice positions, while the loss of outer-shell electrons only results in the formation of electron–hole pairs and does not cause displacement damage, also known as Frenkel defects.

5. Conclusions

Through the above experimental procedure, we investigated the proton and gamma radiation effects on reach-though SPADs. We tested and analyzed the changes in SPAD parameters, such as I–V characteristics, gain and spectral response, before and after irradiation under different bias conditions and proton energies. Based on the analysis and summary of the experimental results, we draw the following conclusions:

1. The leakage current of the SPAD increases with the increase in TID. Under the "$ON_{M=100}$"-bias condition, the increase in the leakage current is significantly larger than that under the "OFF"-bias condition. After 30 krad(Si) gamma irradiation, the leakage current increases to 20.94 nA, resulting in device failure. The change in the leakage current after proton irradiation follows a similar trend as that after gamma irradiation. However, by comparing the parameter degradation laws of protons and gamma rays under the same TID, it is found that the leakage current of the device is more sensitive to the displacement radiation damage caused by protons.
2. Under "OFF"-bias conditions, the gain after gamma irradiation is almost unchanged, meaning that there is no significant change in photocurrent generation. However, after proton irradiation, both the multiplied dark current and the non-multiplied dark current increase significantly compared to before irradiation, leading to an increase in gain.
3. The spectral response after gamma irradiation exhibits fluctuations in the wavelength range of 700–900 nm, while it remains unchanged in other wavelength ranges. When

the proton irradiation fluence reaches 1×10^{11} p/cm^2, the displacement damage defects affect the photoelectron multiplication process, resulting in a decrease in the photocurrent and a decrease in the spectral response.

4. The cross-section for the formation of displacement damage defects by protons in silicon is 1.05×10^5 times greater than the scattering cross-section of gamma rays. The elastic collisions and nuclear reactions between protons and silicon atoms are the main causes of displacement damage in silicon. For SPADs, their parameter degradation is more sensitive to displacement radiation damage.
5. Gamma and proton tests conducted under GTO altitude, a four-year on-orbit mission, and two shielding conditions indicate that the SPAD device will not experience significant performance degradation in leakage current, gain and spectral response parameters. Further testing is required to verify other parameters such as dark count rate (DCR) and photon detection probability (PDP) in these conditions.

Author Contributions: Conceptualization, M.X. and Y.L.; methodology, M.X. and Y.L.; software, M.X.; validation, M.X. and Y.L.; formal analysis, Y.L.; investigation, M.X. and Y.L.; resources, M.X. and M.L.; data curation, M.X.; writing—original draft preparation, M.X.; writing—review and editing, M.X. and Y.L.; visualization, M.X.; supervision, Y.L.; project administration, Y.L.; funding acquisition, Y.L. All authors have read and agreed to the published version of the manuscript.

Funding: This research was funded by The Chinese Academy of Sciences "Western Young Scholars" Program under Grant No. 2021-XBQNXZ-020, Xinjiang "Tianshan Talents" training program of top young talents project under Grant No. 2023TSYCCX0106.

Data Availability Statement: Data are contained within this article.

Conflicts of Interest: The authors declare no conflicts of interest.

References

1. Cusini, I.; Berretta, D.; Conca, E.; Incoronato, A.; Madonini, F.; Maurina, A.A.; Nonne, C.; Riccardo, S.; Villa, F. Historical Perspectives, State of art and Research Trends of Single Photon Avalanche Diodes and Their Applications (Part 1: Single Pixels). *Front. Phys.* **2022**, *10*, 1–12. [CrossRef]
2. Cusini, I.; Berretta, D.; Conca, E.; Incoronato, A.; Madonini, F.; Maurina, A.A.; Nonne, C.; Riccardo, S.; Villa, F. Historical Perspectives, State of Art and Research Trends of SPAD Arrays and Their Applications (Part II: SPAD Arrays). *Front. Phys.* **2022**, *10*, 1–20. [CrossRef]
3. Zhang, J.; Itzler, M.A.; Zbinden, H.; Pan, J.-W. Advances in InGaAs/InP single-photon detector systems for quantum communication. *Light Sci. Appl.* **2015**, *4*, e286. [CrossRef]
4. Tétrault, M.A.; Lamy, É.; Boisvert, A.; Fontaine, R.; Pratte, J.F. Low Dead Time Digital SPAD Readout Architecture for Realtime Small Animal PET. In Proceedings of the 60th IEEE Nuclear Science Symposium (NSS)/Medical Imaging Conference (MIC)/20th International Workshop on Room-Temperature Semiconductor X-ray and Gamma-ray Detectors, Seoul, Republic of Korea, 27 October–2 November 2013.
5. Campajola, M.; Di Capua, F.; Fiore, D.; Nappi, C.; Sarnelli, E.; Gasparini, L. Long-Term Degradation Study of CMOS SPADs in Space Radiation Environment. In Proceedings of the 18th European Conference on Radiation and Its Effects on Components and Systems (RADECS), COBHAM, Goteburg, Sweden, 16–21 September 2018; IEEE: New York, NY, USA, 2018.
6. Haitz, R.H. Model for the Electrical Behavior of a Microplasma. *J. Appl. Phys.* **1964**, *35*, 1370–1376. [CrossRef]
7. Haitz, R.H. Mechanisms Contributing to the Noise Pulse Rate of Avalanche Diodes. *J. Appl. Phys.* **1965**, *36*, 3123–3131. [CrossRef]
8. McIntyre, R.J. A new look at impact ionization—Part I: A theory of gain, noise, breakdown probability, and frequency response. *IEEE Trans. Electron Devices* **1999**, *46*, 1623–1631. [CrossRef]
9. Yuan, P.; Anselm, K.A.; Hu, C.; Nie, H.; Lenox, C.; Holmes, A.L.; Streetman, B.C.; Campbell, J.C.; McIntyre, R.J. A new look at impact ionization—Part II: Gain and noise in short avalanche photodiodes. *IEEE Trans. Electron Devices* **1999**, *46*, 1632–1639. [CrossRef]
10. Zhao, J.X.; Milanese, T.; Gramuglia, F.; Keshavarzian, P.; Tan, S.S.; Tng, M.; Lim, L.; Dhulla, V.; Quek, E.; Lee, M.J.; et al. On Analog Silicon Photomultipliers in Standard 55-nm BCD Technology for LiDAR Applications. *IEEE J. Sel. Top. Quantum Electron.* **2022**, *28*, 10. [CrossRef]
11. Marisaldi, M.; Maccagnani, P.; Moscatelli, F.; Labanti, C.; Fuschino, F.; Prest, M.; Berra, A.; Bolognini, D.; Ghioni, M.; Rech, I.; et al. Single Photon Avalanche Diodes for Space Applications. In Proceedings of the IEEE Nuclear Science Symposium/Medical Imaging Conference (NSS/MIC)/18th International Workshop on Room-Temperature Semiconductor X-ray and Gamma-Ray Detectors, Valencia, Spain, 23–29 October 2011.

12. Wu, M.L.; Ripiccini, E.; Kizilkan, E.; Gramuglia, F.; Keshavarzian, P.; Fenoglio, C.A.; Morimoto, K.; Charbon, E. Radiation Hardness Study of Single-Photon Avalanche Diode for Space and High Energy Physics Applications. *Sensors* **2022**, *22*, 2919. [CrossRef]
13. Vignetti, M.M.; Calmon, F.; Pittet, P.; Pares, G.; Cellier, R.; Quiquerez, L.; Savoy-Navarro, A. Development of a 3D Silicon Coincidence Avalanche Detector for Charged Particle Tracking in Medical Applications. In Proceedings of the IEEE Nuclear Science Symposium/Medical Imaging Conference/Room-Temperature Semiconductor Detector Workshop (NSS/MIC/RTSD), Strasbourg, France, 29 October–6 November 2016; IEEE: New York, NY, USA, 2016.
14. Dauler, E.A.; Grein, M.E.; Kerman, A.J.; Marsili, F.; Miki, S.; Nam, S.W.; Shaw, M.D.; Terai, H.; Verma, V.B.; Yamashita, T. Review of superconducting nanowire single-photon detector system design options and demonstrated performance. *Opt. Eng.* **2014**, *53*, 13. [CrossRef]
15. Gansen, E.J.; Rowe, M.A.; Harrington, S.D.; Nehls, J.M.; Etzel, S.M.; Nam, S.W.; Mirin, R.P. Temperature dependence of the single-photon sensitivity of a quantum dot, optically gated, field-effect transistor. *J. Appl. Phys.* **2013**, *114*, 6. [CrossRef]
16. Vert, A.; Soloviev, S.; Fronheiser, J.; Sandvik, P. Solar-Blind 4H-SiC Single-Photon Avalanche Diode Operating in Geiger Mode. *IEEE Photonics Technol. Lett.* **2008**, *20*, 1587–1589. [CrossRef]
17. Reddy, D.V.; Nerem, R.R.; Nam, S.W.; Mirin, R.P.; Verma, V.B. Superconducting nanowire single-photon detectors with 98% system detection efficiency at 1550 nm. *Optica* **2020**, *7*, 1649–1653. [CrossRef]
18. Natarajan, C.M.; Tanner, M.G.; Hadfield, R.H. Superconducting nanowire single-photon detectors: Physics and applications. *Supercond. Sci. Technol.* **2012**, *25*, 16. [CrossRef]
19. Rosenberg, D.; Kerman, A.J.; Molnar, R.J.; Dauler, E.A. High-speed and high-efficiency superconducting nanowire single photon detector array. *Opt. Express* **2013**, *21*, 1440–1447. [CrossRef] [PubMed]
20. Ni, K.; Zhang, E.X.; Schrimpf, R.D.; Fleetwood, D.M.; Reed, R.A.; Alles, M.L.; Lin, J.Q.; del Alamo, J.A. Gate Bias and Geometry Dependence of Total-Ionizing-Dose Effects in InGaAs Quantum-Well MOSFETs. *IEEE Trans. Nucl. Sci.* **2017**, *64*, 239–244. [CrossRef]
21. Kılıç, A.; Koçak, F.; Piliçer, E.; Tapan, İ. Influence of neutrons on signal fluctuations for APD structures. *Nucl. Instrum. Methods Phys. Res. Sect. A Accel. Spectrometers Detect. Assoc. Equip.* **2013**, *718*, 215–216. [CrossRef]
22. Mohammadnejad, S.; Aasi, M. Analysis of structures and technologies of various types of photodetectors used in laser warning systems: A review. *Opt. Eng.* **2023**, *62*, 090901. [CrossRef]
23. Musienko, Y.; Heering, A.; Ruchti, R.; Wayne, M.; Andreev, Y.; Karneyeu, A.; Postoev, V. Radiation damage in silicon photomultipliers exposed to neutron radiation. *J. Instrum.* **2017**, *12*, 6. [CrossRef]
24. Jouni, A.; Sicre, M.; Malherbe, V.; Mamdy, B.; Thery, T.; Belloir, J.M.; Soussan, D.; De Paoli, S.; Lorquet, V.; Lalucaa, V.; et al. Proton-Induced Displacement Damages in 2-D and Stacked CMOS SPADs: Study of Dark Count Rate Degradation. *IEEE Trans. Nucl. Sci.* **2023**, *70*, 515–522. [CrossRef]
25. Srour, J.R.; Palko, J.W. Displacement Damage Effects in Irradiated Semiconductor Devices. *IEEE Trans. Nucl. Sci.* **2013**, *60*, 1740–1766. [CrossRef]
26. Srour, J.R.; Marshall, C.J.; Marshall, P.W. Review of displacement damage effects in silicon devices. *IEEE Trans. Nucl. Sci.* **2003**, *50*, 653–670. [CrossRef]
27. Pelenitsyn, V.; Korotaev, P. First-principles study of radiation defects in silicon. *Comput. Mater. Sci.* **2022**, *207*, 8. [CrossRef]
28. Wimbauer, T.; Ito, K.; Mochizuki, Y.; Horikawa, M.; Kitano, T.; Brandt, M.S.; Stutzmann, M. Defects in planar Si pn junctions studied with electrically detected magnetic resonance. *Appl. Phys. Lett.* **2000**, *76*, 2280–2282. [CrossRef]
29. Zappa, F.; Tisa, S.; Tosi, A.; Cova, S. Principles and features of single-photon avalanche diode arrays. *Sens. Actuators A Phys.* **2007**, *140*, 103–112. [CrossRef]
30. Dautet, H.; Deschamps, P.; Dion, B.; Macgregor, A.D.; Macsween, D.; McIntyre, R.J.; Trottier, C.; Webb, P.P. Photon-counting techniques with silicon avalanche photodiodes. *Appl. Opt.* **1993**, *32*, 3894–3900. [CrossRef] [PubMed]
31. Cova, S.; Ghioni, M.; Lotito, A.; Rech, I.; Zappa, F. Evolution and prospects for single-photon avalanche diodes and quenching circuits. *J. Mod. Opt.* **2004**, *51*, 1267–1288. [CrossRef]
32. Hadfield, R.H. Single-photon detectors for optical quantum information applications. *Nat. Photonics* **2009**, *3*, 696–705. [CrossRef]
33. Allison, J.; Amako, K.; Apostolakis, J.; Arce, P.; Asai, M.; Aso, T.; Bagli, E.; Bagulya, A.; Banerjee, S.; Barrand, G.; et al. Recent developments in GEANT4. *Nucl. Instrum. Methods Phys. Res. Sect. A* **2016**, *835*, 186–225. [CrossRef]
34. Allison, J.; Amako, K.; Apostolakis, J.; Araujo, H.; Dubois, P.A.; Asai, M.; Barrand, G.; Capra, R.; Chauvie, S.; Chytracek, R.; et al. Geant4 developments and applications. *IEEE Trans. Nucl. Sci.* **2006**, *53*, 270–278. [CrossRef]
35. Agostinelli, S.; Allison, J.; Amako, K.; Apostolakis, J.; Araujo, H.; Arce, P.; Asai, M.; Axen, D.; Banerjee, S.; Barrand, G.; et al. GEANT4-a simulation toolkit. *Nucl. Instrum. Methods Phys. Res. Sect. A* **2003**, *506*, 250–303. [CrossRef]
36. Akkerman, A.; Barak, J.; Chadwick, M.B.; Levinson, J.; Murat, M.; Lifshitz, Y. Updated NIEL calculations for estimating the damage induced by particles and γ-rays in Si and GaAs. *Radiat. Phys. Chem.* **2001**, *62*, 301–310. [CrossRef]
37. Xun, M.; Li, Y.; Feng, J.; He, C.; Liu, M.; Guo, Q.; Caputo, D. Effect of Proton Irradiation on Complementary Metal Oxide Semiconductor (CMOS) Single-Photon Avalanche Diodes. *Electronics* **2024**, *13*, 224. [CrossRef]
38. Ning, B.; Zhang, Z.; Liu, Z.; Hu, Z.; Chen, M.; Bi, D.; Zou, S. Radiation-induced shallow trench isolation leakage in 180-nm flash memory technology. *Microelectron. Reliab.* **2012**, *52*, 130–136. [CrossRef]

39. Osborne, M.D.; Hobson, P.R.; Watts, S.J. Numerical simulation of neutron radiation effects in avalanche photodiodes. *IEEE Trans. Electron Devices* **2000**, *47*, 529–536. [CrossRef]
40. Moscatelli, F.; Marisaldi, M.; Maccagnani, P.; Labanti, C.; Fuschino, F.; Prest, M.; Berra, A.; Bolognini, D.; Ghioni, M.; Rech, I.; et al. Radiation tests of single photon avalanche diode for space applications. *Nucl. Instrum. Methods Phys. Res. Sect. A* **2013**, *711*, 65–72. [CrossRef]
41. Bychkova, O.; Garutti, E.; Popova, E.; Stifutkin, A.; Martens, S.; Parygin, P.; Kaminsky, A.; Schwandt, J. Radiation damage uniformity in a SiPM. *Nucl. Instrum. Methods Phys. Res. Sect. A* **2022**, *1039*, 4. [CrossRef]
42. Watts, S.J. Overview of radiation damage in silicon detectors—Models and defect engineering. *Nucl. Instrum. Methods Phys. Res. Sect. A* **1997**, *386*, 149–155. [CrossRef]
43. Ratti, L.; Brogi, P.; Collazuol, G.; Betta, G.F.D.; Ficorella, A.; Marrocchesi, P.S.; Morsani, F.; Pancheri, L.; Torilla, G.; Vacchi, C. DCR Performance in Neutron-Irradiated CMOS SPADs From 150-to 180-nm Technologies. *IEEE Trans. Nucl. Sci.* **2020**, *67*, 1293–1301. [CrossRef]
44. Ziegler, J.F.; Ziegler, M.D.; Biersack, J.P. SRIM—The stopping and range of ions in matter (2010). *Nucl. Instrum. Methods Phys. Res. Sect. B Beam Interact. Mater. At.* **2010**, *268*, 1818–1823. [CrossRef]
45. Liu, Q.L.; Zhang, H.Y.; Hao, L.X.; Hu, A.Q.; Wu, G.; Guo, X. Total dose test with γ-ray for silicon single photon avalanche diodes*. *Chin. Phys. B* **2020**, *29*, 088501. [CrossRef]
46. Ratti, L.; Brogi, P.; Collazuol, G.; Dalla Betta, G.F.; Ficorella, A.; Lodola, L.; Marrocchesi, P.S.; Mattiazzo, S.; Morsani, F.; Musacci, M.; et al. Dark Count Rate Degradation in CMOS SPADs Exposed to X-Rays and Neutrons. *IEEE Trans. Nucl. Sci.* **2019**, *66*, 567–574. [CrossRef]
47. Summers, G.P.; Burke, E.A.; Shapiro, P.; Messenger, S.R.; Walters, R.J. Damage correlations in semiconductors exposed to gamma-radiation, electron-radiation and proton-radiation. *IEEE Trans. Nucl. Sci.* **1993**, *40*, 1372–1379. [CrossRef]

Disclaimer/Publisher's Note: The statements, opinions and data contained in all publications are solely those of the individual author(s) and contributor(s) and not of MDPI and/or the editor(s). MDPI and/or the editor(s) disclaim responsibility for any injury to people or property resulting from any ideas, methods, instructions or products referred to in the content.

electronics

MDPI

Article

Evaluation and Mitigation of Weight-Related Single Event Upsets in a Convolutional Neural Network

Yulong Cai [1], Ming Cai [2], Yanlai Wu [1], Jian Lu [1], Zeyu Bian [1], Bingkai Liu [3,*] and Shuai Cui [1,*]

1 Innovation Academy for Microsatellites of Chinese Academy of Sciences, Shanghai 200020, China; caiyl@microsate.com (Y.C.)
2 University of Chinese Academy of Sciences, Beijing 100049, China; caiming@iie.ac.cn
3 Key Laboratory of Functional Materials and Devices for Special Environments, Xinjiang Technical Institute of Physics and Chemistry, Chinese Academy of Sciences, Urumqi 830011, China
* Correspondence: liubk@ms.xjb.ac.cn (B.L.); cuis@microsate.com (S.C.)

Abstract: Single Event Upsets (SEUs) are most likely to cause bit flips within the trained parameters of a convolutional neural network (CNN). Therefore, it is crucial to analyze and implement hardening techniques to enhance their reliability under radiation. In this paper, random fault injections into the weights of LeNet-5 were carried out in order to evaluate and propose strategies to improve the reliability of a CNN. According to the results of an SEU fault injection, the accuracy of the CNN can be classified into the following three categories: benign conditions, poor conditions, and critical conditions. Two efficient methods for mitigating weight-related SEUs are proposed, as follows: weight limiting and Triple Modular Redundancy (TMR) for the critical bit of the critical layer. The hardening results show that when the number of SEU faults is small, the weight limiting almost completely eliminates the critical and poor conditions of LeNet-5's accuracy. Additionally, even when the number of SEU faults is large enough, combining the weight limiting and TMR methods for the critical bit of the critical layer can retain the occurrence rate of benign conditions at 98%, saving 99.3% of the hardware resources compared to the Full-TMR hardening method.

Keywords: reliability; SEU; CNN; weight

1. Introduction

Today, convolutional neural networks (CNNs) are adopted in various domains that range from high-performance computing to data analysis and target recognition for space applications [1–4]. However, when applied in the spaceflight field, CNNs face reliability challenges due to the impact of space radiation. This is because the memory of the hardware accelerators carrying CNNs have been proven to be highly sensitive to energetic particle attacks, especially Single Event Upset (SEU) effects [5–8]. The memory of hardware accelerators stores a large number of weight parameters for CNNs. As process technology scales up, the area consumed by memory dominates the CNN hardware accelerator's area. Consequently, SEUs will become a major reliability concern for future CNN accelerators. Weight-related SEUs will further propagate in CNN computations, ultimately leading to a decrease in CNN accuracy.

Therefore, it is necessary to conduct an assessment of the impact of weight-related SEUs on CNNs and design appropriate hardening measures. Clarifying the fault modes and mechanisms of weight-related SEUs causing CNN accuracy degradation is the basis for studying hardening methods. In [9,10], weight-related SEUs in a CNN were divided into tolerable faults and critical faults, with tolerable faults sometimes referred to as Silent Data Corruption (SDC) [11]. This paper further explores the relationship between weight-related SEUs and the classification accuracy of CNNs, categorizing the impact of weight-related SEUs on CNN accuracy into the following three categories: benign conditions, poor conditions, and critical conditions.

There are many traditional system-level SEU mitigation methods [12,13]; however, due to the large computational requirements and numerous parameters of CNNs, these system mitigation methods are not well suited for CNN hardening needs. On the one hand, the significant number of weight parameters in CNNs occupies a large portion of the storage resources in onboard hardware accelerators, causing the Full-TMR methods to exacerbate the already limited hardware resources. Additionally, error correction coding methods can reduce the CNN's computational speed, while being unable to correct multiple SEU errors.

To overcome the limitations of traditional hardening methods, research on CNN hardening is primarily based on two main approaches. Firstly, it involves improved methods based on the traditional hardening strategies. TMR is a typical method; however, the overhead of TMR can be very high and it can jeopardize the device's performance and efficiency [14,15]. Since not all the errors are critical for a CNN, it is crucial to accurately identify the critical layers of the CNN to balance the utilization of the TMR hardware resources and the hardening effectiveness. In [16], they tested the robustness of the CNN architecture via injecting Gaussian noise into the weights of the CNN and concluded that the error tolerance of a layer tended to worsen as it moved toward the output layer. However, the sensitivity of weight-related SEUs in different layers is influenced by various factors such as the number of weights, the weight value distribution, the network structure, and the layer position. Therefore, this paper characterizes the SEU sensitivity of the weights of different layers in a CNN. Secondly, it involves optimizing the design of CNN algorithms. In [17,18], the method of weight quantization can reduce the sensitivity of CNNs to SEUs. In [19], three modern deep convolutional CNNs were tested for their robustness. It was hypothesized that the use of batch normalization or shortcut connections made ResNet50 and InceptionV3 resilient against memory errors. Some interesting works have proposed adapting the existing Algorithm-Based Fault-Tolerant (ABFT) [20]. However, there is currently no algorithmic optimization hardening method that can significantly improve the SEU resilience of CNN weights.

The main innovations of this paper can be summarized in three points, as follows: firstly, discovering the uncertainty caused by an SEU and classifying the fault modes, with particular emphasis on the possibility of severe consequences, even with a low number of faults; secondly, pointing out that weight-limiting hardening can effectively address this issue, while also improving the rate of benign conditions; and thirdly, proposing a method for identifying critical layers and critical bits and demonstrating the cumulative effects of TMR and weight-limiting hardening. The structure of the remaining sections of this paper is as follows: Section 2 provides an introduction to the LeNet-5 CNN, along with the SEU fault injection method and the simulation results before hardening. In Section 3, we delve into the weight limiting and TMR methods for the critical bit of the critical layer and present their associated outcomes. Finally, Section 4 offers concluding remarks and sets the stage for future work.

2. Fault Injection

2.1. Simulation Framework

The neural network model used in this work is LeNet-5 and the network structure is shown in Figure 1. We used the Modified National Institute of Standards and Technology (MNIST) dataset, which is composed of hand-written digits, as the input for the LeNet-5 CNN test. The MNIST is commonly used for training various image processing systems or testing in the fields of machine learning.

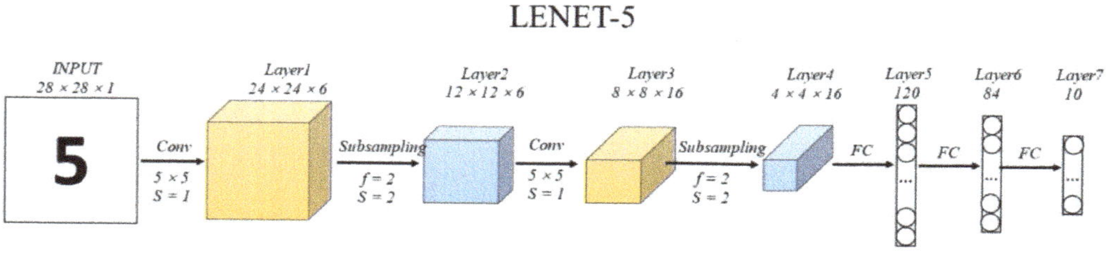

Figure 1. The structure of LeNet-5.

Figure 2 represents the framework of simulation. Firstly, the SEUs were introduced to the LeNet-5 after training was complete. Then, the LeNet-5's accuracy was re-evaluated with the new weight. Each weight is represented by 32-bit floating point numbers. The number of bits in SEUs were increased to determine how susceptible the LeNet-5 was to weight-related SEUs and to classify the behavior of the LeNet-5 under an increasing number of SEUs. The bits in SEUs were increased from 0 to 225 and the classification accuracy of LeNet-5 was recorded. In total, 50 trials were recorded for each number of bits in SEUs.

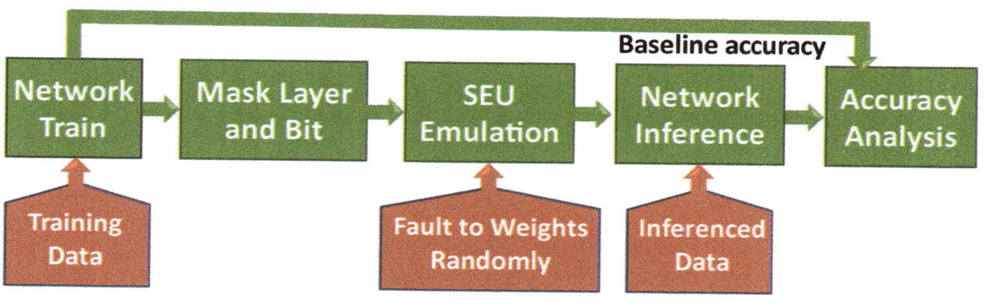

Figure 2. Framework for quantifying CNN degradations caused by injecting fault to weights randomly.

The pseudocode for SEU fault injections is shown in the pseudocode Algorithm 1; the theoretical flip rate for each bit is first calculated based on the bit flip rate of the weight storage unit and the duration of the on-orbit operation. Then, all weight bits are traversed, each being flipped with a probability of p. The number of SEU faults injected is a statistic obtained by multiplying the SEU rate by the total number of weights. To simulate the real-world scenario where the same bit can be flipped multiple times, the on-orbit duration is divided into batch-size parts, with faults being injected for a duration of years/batch size each time. Additionally, the simulation models the situation where multiple bit flips are caused by the accumulation of SEUs; however, it does not simulate the scenario where multiple bit flips are caused by a single particle radiation. Furthermore, TMR (Triple Modular Redundancy) and weight limit hardening have been set as adjustable options, making it convenient to obtain the effects of different hardening combinations. The purpose of this is to provide more flexibility, allowing the most suitable hardening strategy to be chosen based on different requirements and conditions.

Algorithm 1 Inject-Fault

```
Input: mod-el, datatype, TMR, p, years, limitvalid, batchsize
Output: model
/* model: LENET-5 model.                                                   */
/* datatype: Data type for model's weights and bias.                       */
/* TMR: Zero-TMR, Partial-TMR and Full-TMR.                                */
/* p: The probability of flipping a single bit per year.                   */
/* years: The duration of exposure for devices in the radiation space.     */
/* limitvalid: Flag for limiting the weights after fault injection.        */
/* batchsize: Inject faults in multiple batches.                           */

weights=model.GetWeights()
if datatype == "float32" then
    for t←0 to years/batchsize do
        p = batchsize * p
        for index ← 0 to len(weights) do
            data = float32_to_bin(weight[index])
            for dindex ← 0 to len(data) do
                if NoTMR(TMR, dindex) then
                    //the function SeuRandom is used to generate an enable signal
                        with a probability of p being effective
                    Flip(data[dindex], SeuRandom(p))
                else
                    flip(data[dindex], SeuRandom(PTMR(p)))
            Weights[index] = bin_to_float32(data)
            // the function isnan is used to check if a number is invalid
            if isnan(weights[index]) then
                weights[index] = 0
            else
                if limitvalid and (weights[index]>uplimit) then
                    weights[index] = weights.max() >> 1
                else if limitvalid and (weights[index]<lowlinit) then
                    Weights[index] = weights.min() >> 1
else
    // operations of other datatype similar to float32
    ...
Model.SetWeights(reshape(weights))
```

2.2. Results before Hardening

Before hardening, the accuracy of LeNet-5 varies with the number of SEU fault injections, as shown in Figure 3. Each column of 50 blue dots represents 50 trials, with brighter colors indicating a higher frequency of occurrences. From the graph, the following three phenomena can be observed: First, despite the same number of SEU fault injections, there is a significant difference in LeNet-5's accuracy. Second, there is a high frequency of extremely low LeNet-5 accuracy (below 0.3), even when the number of SEU fault injections is small, with a 31% probability of critical situations occurring. Third, as the number of SEU fault injections increases, the maximum accuracy of LeNet-5 rapidly decreases and, when the number of SEU fault injections reaches 125, almost all classification accuracies are below 0.85. Next, we will delve into the reasons behind these three phenomena, propose corresponding efficient hardening methods, and validate the effectiveness of the hardening measures through SEU fault simulation.

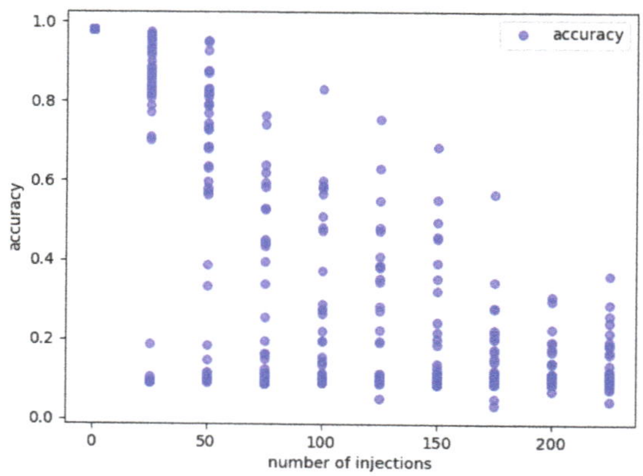

Figure 3. LeNet-5's accuracy varies with the number of SEUs injected.

3. Hardening Methods and Validation

Based on the changes in CNN accuracy and the acceptance criteria for specific spaceflight tasks, we categorize the SEU-induced CNN accuracy into the following three categories: benign, poor, and critical. The benign condition refers to an accuracy higher than the acceptable accuracy for spaceflight tasks. In this paper, we define an accuracy greater than 85% as a benign condition. The poor condition refers to an accuracy between the accuracy of blind guessing and the acceptable accuracy for spaceflight tasks. In this paper, we define an accuracy between 30% and 85% as a poor condition. The critical condition refers to an accuracy lower than the accuracy of blind guessing. Figure 4 illustrates the relationship between the three types of LeNet-5 accuracy and the number of SEU fault injections in the absence of hardening. With an increase in the number of SEU faults, the benign condition gradually decreases. When the number of SEU faults reaches around 50, the poor condition has the highest proportion. When the number of faults increases to 75, the critical condition already accounts for 70% of cases, indicating that LeNet-5 is almost unable to function properly. To improve CNN accuracy, the primary goal is to minimize the occurrence of critical conditions. It is important to note that the CNN accuracy classification method provided in this manuscript is universal, but the threshold conditions and spaceflight task requirements for different classifications (benign, poor, and critical) are closely related. Different task requirements will affect the threshold values. In our engineering code, the threshold for determining "benign" and "poor" conditions is designed as a variable.

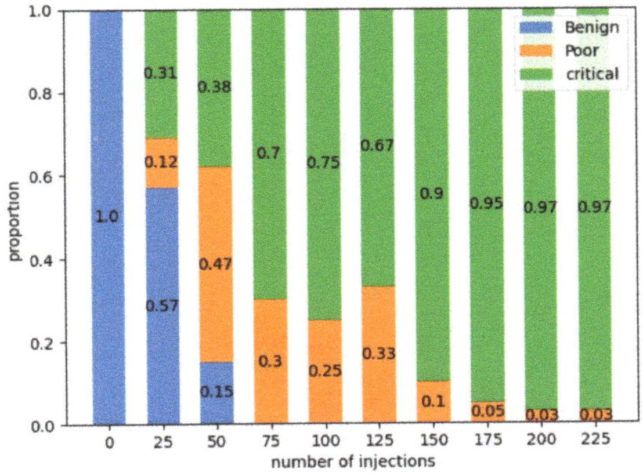

Figure 4. The proportion of the three LeNet-5 accuracy conditions caused by weight-related SEUs.

The statistical fault injection method in a CNN was discussed [21], and the fault injection results of this project were statistically analyzed under experimental conditions. The abscissa in Figure 4 represents the number of fault injections. We will separately analyze the distribution of the network model for different numbers of fault injections to reflect the dynamic changes of the network in orbit. After specifying the number of fault injections x, the sample population of fault injections becomes all combinations of selecting x bits from all weight bits. In our experiment, the sample size was selected as 50. For each combination, when the selected weights are flipped, the CNN inferences a test set of 10 k images and obtains a final accuracy. Then, based on the accuracy, the CNN output situation is classified (for each specific classification, it is a binomial distribution). Therefore, in the case of sampling sample size n, the standard error of the proportion of one output situation of CNN is:

$$\mathrm{SE}(\mathrm{p}) = \sqrt{\frac{\mathrm{p} \times (1-\mathrm{p})}{\mathrm{n}}} \tag{1}$$

where p represents the proportion of one output situation of the CNN in the sampled sample. The confidence interval is:

$$\mathrm{p} \pm \mathrm{z} \times \mathrm{SE}(\mathrm{p}) \tag{2}$$

where z and confidence level are related. According to Figure 4, when the injection count is 25 (benign situation accounts for 0.57), the sample size is 50, the confidence level is 90%, the standard error of the benign situation is 0.07, and the confidence interval is [0.45, 0.68]. The errors and confidence interval conditions of the results at other injection counts are even better than this. We believe that such results are acceptable.

3.1. Weight Limiting

Research has shown that the distribution of weights in CNNs follows an approximately normal distribution with a mean of 0, indicating that the majority of weight values are small [22]. The weight distributions of the LeNet-5 network are shown in Figures 5 and 6, where Figure 5 represents the weight distribution of the second convolutional layer and Figure 6 represents the weight distribution of the first fully connected layer. Similar patterns can be observed in the weight distributions of other convolutional and fully connected layers.

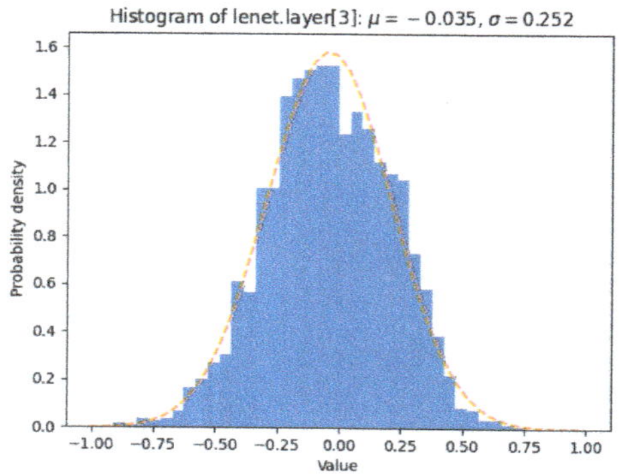

Figure 5. The weight distributions of the third layer of the LeNet-5 network.

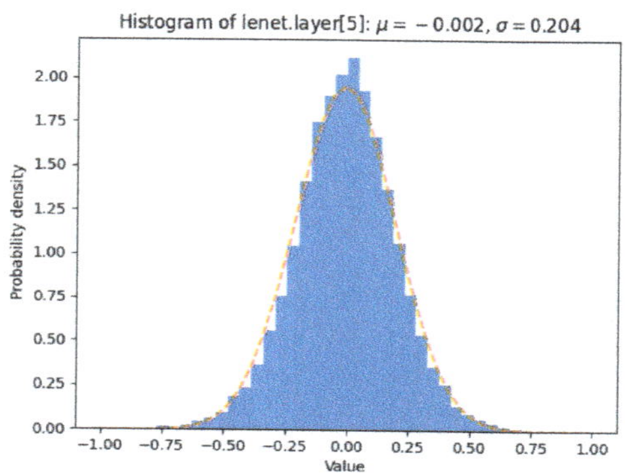

Figure 6. The weight distributions of the sixth layer of the LeNet-5 network.

Therefore, weight-related SEUs can potentially cause extreme weight values and, in this study, weights that exceed twice the normal weight value are defined as extreme weights. Extreme weights can have a significant impact on the layer in which they are present; when extreme weights occur in the convolutional kernels of a convolutional layer, the other weights in that kernel no longer function, leading to the failure of an output channel in that convolutional layer. Similarly, when extreme weight values occur in a specific row of the weight matrix in a fully connected layer, the corresponding output neuron in that row becomes ineffective. Moreover, errors caused by the changing weights propagate backward through the network, so even a small number of extreme weights can cause the entire network to degrade into a critical condition.

Next, we will establish a model to quantitatively evaluate the probability of extreme weights, leading to the degradation of CNN accuracy into a critical condition after T years. This model primarily considers the following two factors: the probability of SEUs causing weights to become extreme and the probability of extreme weights leading to a critical condition.

The probability p1 is defined as the actual SEU probability of a bit storing weights in the hardware accelerator. Considering that a bit can be affected by multiple SEUs, we can calculate the actual SEU probability p2 of a bit after T years using Formula (3), and the probability p3 represents that a weight has flip bit (≥1 bit), where bitlen represents the bit width of the data type:

$$\mathrm{p2} = \sum_{2\nmid \mathrm{d}, \mathrm{d} \leq \mathrm{T}} \mathrm{C}_{\mathrm{T}}^{\mathrm{d}} \mathrm{p1}^{\mathrm{d}} (1 - \mathrm{p1})^{\mathrm{T}-\mathrm{d}} \tag{3}$$

$$\mathrm{p3} = 1 - \mathrm{C}_{\mathrm{bitlen}}^{0} \times (1 - \mathrm{p2})^{\mathrm{bitlen}} \tag{4}$$

Then, we analyze the probability of SEUs causing a weight to become extreme. The weights discussed in this study are represented as single-precision floating-point numbers, which consist of three parts, as follows: sign bit, exponent, and mantissa. Once a high-order bit of the exponent is affected by an SEU, it can have a significant impact on the value of the floating-point number (IEEE 754), as shown in the example in Figure 7. To obtain the probability of a weight becoming extreme, we first generate a set of data that follows a normal distribution, based on the mean and standard deviation of the weights. Then, we calculate the probability that each number in this set becomes an extreme weight due to SEUs and take the arithmetic average of these probabilities. The resulting probability is 6.7%, which we denote as p4.

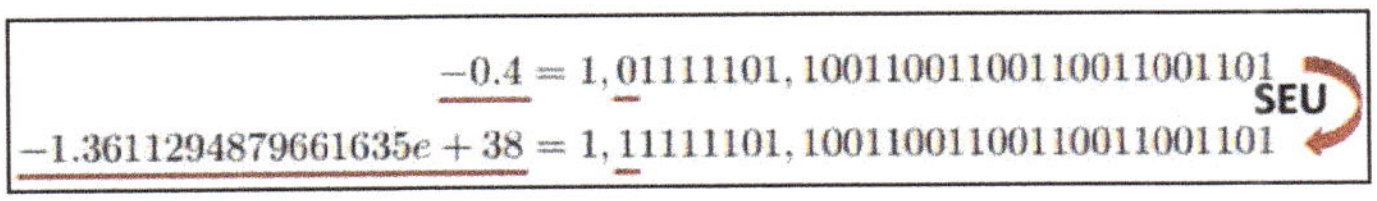

Figure 7. Example of an SEU causing an extreme weight.

p5[x] represents the probability that the number of extreme weights is x after T years. p6[x] represents the probability of the model becoming a critical condition when there are x extreme weights in the CNN.

$$\mathrm{p5[x]} = \mathrm{C_N^x}(\mathrm{p3} * \mathrm{p4})^{\mathrm{x}}(1 - \mathrm{p3} * \mathrm{p4})^{\mathrm{N-x}} \tag{5}$$

$$\mathrm{p6[x]} = \frac{\sum_{\mathrm{i=0}}^{\mathrm{n-1}} \mathrm{flag\_iter[i]}}{\mathrm{n}}, \mathrm{flag\_iter[i]} \in \{0, 1\} \tag{6}$$

In the equation, flag_iter[i] represents whether the CNN test result of the i-th iteration is of a critical condition and n represents the number of random injection tests performed for each number of extreme weights. p represents the probability that the CNN reaches a critical condition in the Tth year:

$$\mathrm{p} = \sum_{\mathrm{x=0}}^{\mathrm{N}} \mathrm{p5[x]p6[x]} \tag{7}$$

Through Formula (7), we found that when no hardening is performed, the probability of extreme weights causing LeNet-5's accuracy to reach a critical condition is as high as 99% when the number of SEU fault injections reaches 225. Therefore, we can conclude that 6.7% of SEUs dominate the 97% critical condition in Figure 2, while the remaining 93.3% of SEUs account for the 0.03% poor condition. Additionally, when the number of SEUs is below a certain value, both poor conditions and critical conditions are mainly caused by extreme weights.

Based on the characteristics of weight distribution and the main causes of LeNet-5's accuracy degradation, we propose a new hardening method, weight limiting. The principle is shown in the following equation:

$$\mathrm{w_{buffer}} = \begin{cases} 0 & \mathrm{w\_men} \in \{\mathrm{nan,\ inf}\} \\ \mathrm{wmax} \gg 1 & \mathrm{w\_mem} \leq \mathrm{uplimit} \\ \mathrm{wmin} \gg 1 & \mathrm{w\_mem} \geq \mathrm{lowlimit} \end{cases} \tag{8}$$

In the equation, w_buffer represents the weights in the buffer closest to the calculation unit, w_mem represents the weights stored at the previous level of w_buffer, nan (not a number) and inf (infinite) are used to determine if the current weights are invalid, and uplimit and lowlimit are the upper and lower limits, respectively. The variables wmax and wmin represent the maximum and minimum weight values, respectively. The symbol ">>" represents a right shift of 1 bit. This means that the binary representation of a number is shifted to the right by one position, effectively dividing the number by 2.

In Figures 8 and 9, weight limiting and Full-TMR are compared as two hardening methods. Figure 8 compares the probability of LeNet-5's accuracy reaching a critical condition, while Figure 9 compares the probability of LeNet-5's accuracy reaching a benign condition. The results in Figure 8 indicate that the weight limiting method can effectively eliminate critical conditions, as extreme weights are the main cause of critical conditions. From Figure 9, it can be observed that when the number of SEUs is less than 380, weight limiting significantly improves the proportion of LeNet-5's benign accuracy, reaching as high as 98%. So extreme weights are also the dominant cause of poor conditions at this stage. As the number of SEUs further increases, the probability of critical conditions remains close to 0% under the weight limiting, but the benign condition starts to degrade. This is because the limiting measures cannot eliminate the small range fluctuations in weights caused by SEUs. Moreover, with an increasing number of SEU faults, the accumulation of small range fluctuations in weights gradually has a greater impact on LeNet-5's accuracy.

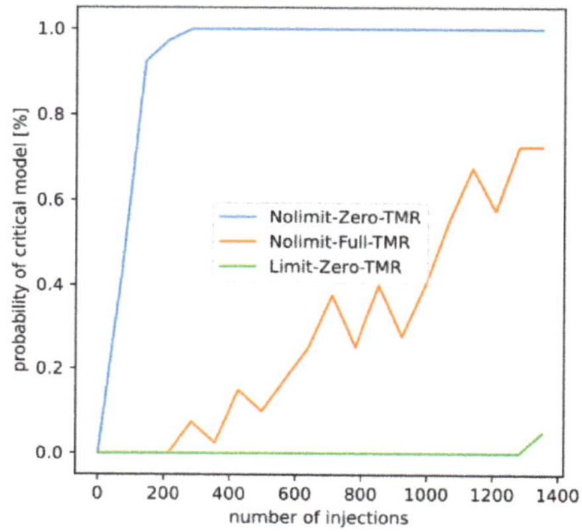

Figure 8. The probabilities of critical condition change with the number of SEUs under different hardening methods.

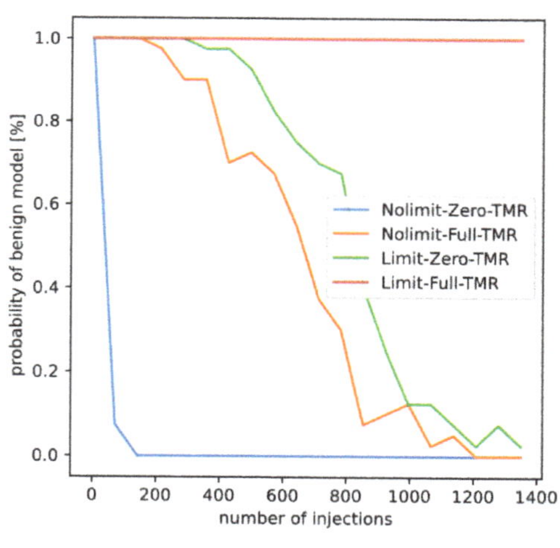

Figure 9. The probabilities of benign condition change with the number of SEUs under different hardening methods.

Figure 9 shows that when weight limiting is combined with Full-TMR, it can effectively address the issue of declining benign conditions. However, the Full-TMR method consumes a large amount of hardware resources and is not suitable for scenarios with a large number of weights, such as CNNs. Therefore, it becomes particularly important to select key positions for Partial-TMR.

3.2. Selective TMR

The TMR process for the critical bits of CNN weights can be divided into two steps. The first step is to identify the critical weight layers. The second step is to harden the critical layers by protecting the sign bit, the most significant bit of the exponent, and the most significant bit of the mantissa of the weights in those layers. It is particularly important to note that the weight limiting method has already been applied to LeNet-5 in the following simulation.

To identify the critical layers of LeNet-5, first, TMR is applied to the weights of different layers individually. Then, SEU faults are randomly injected into the weights of all layers; the accuracy results of LeNet-5 are shown in Figure 10. Figure 10 shows the results of TMR hardening applied to the weights of each layer in LeNet-5 and there is no apparent dependency on the effectiveness of TMR across different layers. Considering the number of weights in each layer, as given in Table 1, the first layer, the third layer, and the seventh layer show the highest efficiency in hardening. This means that significant improvement in LeNet-5's accuracy can be achieved by applying TMR to a small number of weights in these three layers. Therefore, these three layers are defined as the critical layers. Figure 11 compares the hardening results of all layers, critical layers, convolutional layers, and fully connected layers. The results show that the TMR of the critical layers has a significantly better effect than the convolutional layers and fully connected layers.

Next, we compare the following five scenarios: TMR for full-bit of all layers (all-all-TMR), TMR for full-bit of the critical layer(partial-all-TMR), TMR for the critical bit of all layers (all-partial-TMR), TMR for the critical bit of the critical layer (partial-partial-TMR), and no TMR (Zero-TMR). The simulation results are shown in Figure 12. It is evident that the best effect is achieved with TMR for the full-bit of all layers, followed by the full-bit of the critical layer (92.3% reduction in resources) and the critical bit of all layers (90.6% reduction in resources). Next is the critical bit of the critical layer (99.3% reduction in resources) and, finally, no TMR (100% reduction in resources).

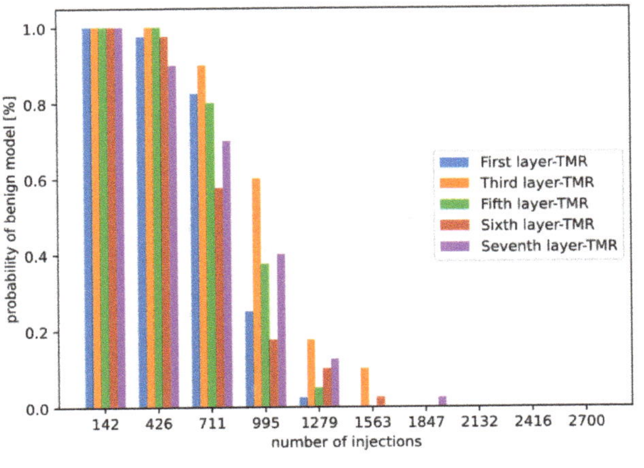

Figure 10. The probability of benign conditions after hardening for individual layers in LeNet-5.

Table 1. The exact number of weights in each layer of LeNet-5.

Layer	All	Layer 1	Layer 2	Layer 3	Layer 4	Layer 5	Layer 6	Layer 7
Number of weights	44,426	156	0	2416	0	30,840	10,164	850

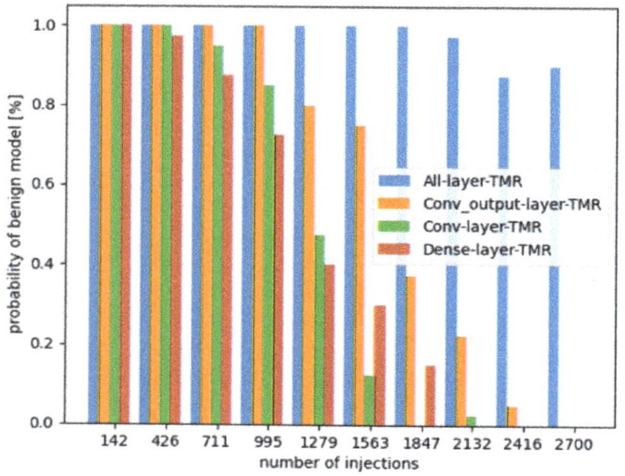

Figure 11. The probability of benign conditions after hardening for the combination of multiple layers in LeNet-5.

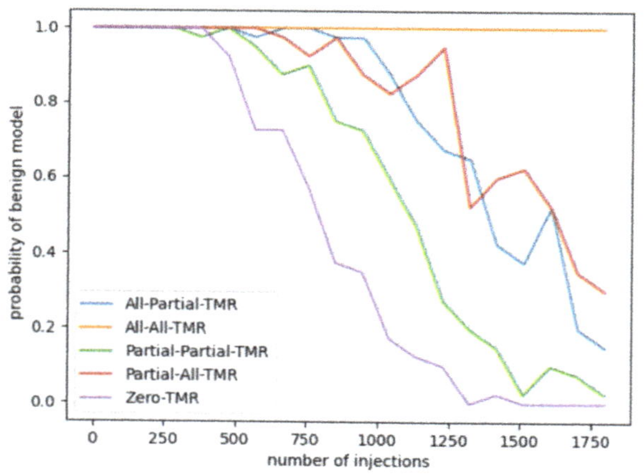

Figure 12. Comparison of the effectiveness of five different selective TMR methods.

The choice of the final hardening method depends on factors such as the mission environment, radiation environment, on-orbit duration, device sensitivity, CNN accuracy classification criteria, network model, dataset, hardware resources, and task requirements. Under the given conditions in this paper, when the number of on-orbit SEU fault injections is less than 600, selecting the partial-partial-TMR hardening method (weight limiting method is already applied by default) can maintain the benign condition probability of LeNet-5 at 98%.

4. Conclusions

In this paper, it was discovered that impact of weight-related SEUs on CNN accuracy has significant uncertainty and, even at very low SEU fault injection rates, extremely low accuracy can occur. This phenomenon indicates that the CNN accuracy obtained from a single irradiation experiment cannot represent its true accuracy. Based on CNN accuracy and specific space mission requirements, we classify the CNN accuracy into the following three categories: benign conditions, poor conditions, and critical conditions.

This paper analyzes the reasons for the degradation in CNN accuracy from two aspects. First, SEUs lead to extreme weights, which significantly affect the layer where the weights are located. Second, SEUs cause small variations in weight values and, only when a certain number of such weights accumulate, can they lead to poor or critical CNN conditions. We establish a mathematical model to quantitatively evaluate the probability of extreme weights causing CNN accuracy to degrade into a critical condition. The model shows that the probability of SEUs causing weights to become extreme is 6.7%. When the number of SEU fault injections reaches 225, the probability of extreme weights leading to CNN accuracy reaching critical condition reaches 99%.

Two hardening methods are proposed, as follows: weight limiting and partial-partial-TMR. When the number of SEU faults is below a certain threshold, weight limiting reduces the probabilities of critical and poor conditions in LeNet-5 to almost zero, because extreme weights dominate the accuracy degradation. As the number of SEU faults increases, the accumulation of small variations in weights starts to affect the benign condition. We identify the critical layers of LeNet-5 as the input for Selective TMR, which are the first layer, third layer, and seventh layer. Under the given application conditions in this paper, when the number of SEU faults is large, both weight limiting and partial-partial-TMR methods maintain a benign condition occurrence rate of approximately 98%, while reducing TMR resource consumption by 99.3%. It is important to note that the quantitative analysis conclusions in this paper may vary depending on the CNN type, accuracy classification criteria, and definition of extreme weights. However, the qualitative analysis conclusions apply to all CNN weight flips.

Author Contributions: Conceptualization, Y.C.; methodology, S.C. and B.L.; software, M.C.; validation, Y.W.; formal analysis, Z.B.; writing—original draft preparation, J.L.; writing—review and editing, Y.C. All authors have read and agreed to the published version of the manuscript.

Funding: This research was funded by the National Natural Science Foundation of China, grant number 12305323" and, in part, by the Innovation Foundation of Radiation Application, China Institute of Atomic Energy, under Grant KFZC2022020101.

Data Availability Statement: The original data presented in the study are openly available in GitHub at https://github.com/minghan1/Letnet-seu.git, accessed on 1 March 2023.

Conflicts of Interest: The authors declare no conflicts of interest.

References

1. Motamedi, M.; Gysel, P.; Akella, V.; Ghiasi, S. Design space exploration of FPGA-based Deep Convolutional CNNs. In Proceedings of the 2016 21st Asia and South Pacific Design Automation Conference (ASP-DAC), Macao, China, 25–28 January 2016; pp. 575–580.
2. Wu, T.; Yang, X.; Song, B.; Wang, N. T-SCNN: A Two-Stage Convolutional CNN for Space Target Recognition. In Proceedings of the IGARSS 201-2019 IEEE International Geoscience and Remote Sensing Symposium, Yokohama, Japan, 28 July–2 August 2019; pp. 1334–1337.
3. Zou, A.-M.; Kumar, K.D.; Hou, Z.-G. Quaternion-based adaptive output feedback attitude control of spacecraft using chebyshev neural networks. *IEEE Trans. Neural Networks* **2010**, *21*, 1457–1471. [CrossRef]
4. Gao, H.; Cheng, B.; Wang, J.; Li, K.; Zhao, J.; Li, D. Object classification using CNN-based fusion of vision and LIDAR in autonomous vehicle environment. *IEEE Trans. Ind. Inform.* **2018**, *14*, 4224–4231. [CrossRef]
5. Sierawski, B.D.; Pellish, J.A.; Reed, R.A.; Schrimpf, R.D.; Warren, K.M.; Weller, R.A.; Mendenhall, M.H.; Black, J.D.; Tipton, A.D.; Xapsos, M.A.; et al. Impact of Low-Energy Proton Induced Upsets on Test Methods and Rate Predictions. *IEEE Trans. Nucl. Sci.* **2009**, *56*, 3085–3092. [CrossRef]

6. Warren, K.; Weller, R.; Mendenhall, M.; Reed, R.; Ball, D.; Howe, C.; Olson, B.; Alles, M.; Massengill, L.; Schrimpf, R.; et al. The contribution of nuclear reactions to heavy ion single event upset cross-section measurements in a high-density SEU hardened SRAM. *IEEE Trans. Nucl. Sci.* **2005**, *52*, 2125–2131. [CrossRef]
7. Su, F.; Liu, C.; Stratigopoulos, H.-G. Testability and dependability of AI hardware: Survey, trends, challenges, and perspectives. *IEEE Des. Test* **2023**, *40*, 8–58. [CrossRef]
8. Ibrahim, Y.; Wang, H.; Liu, J.; Wei, J.; Chen, L.; Rech, P.; Adam, K.; Guo, G. Soft errors in DNN accelerators: A comprehensive review. *Microelectron. Reliab.* **2020**, *115*, 113969. [CrossRef]
9. Santos, F.F.D.; Draghetti, L.; Weigel, L.; Carro, L.; Navaux, P.; Rech, P. Evaluation and Mitigation of Soft-Errors in CNN-Based Object Detection in Three GPU Architectures. In Proceedings of the 2017 47th Annual IEEE/IFIP International Conference on Dependable Systems and CNNs Workshops (DSN-W), Denver, CO, USA, 26–29 June 2017; pp. 169–176.
10. Libano, F.; Wilson, B.; Anderson, J.; Wirthlin, M.; Cazzaniga, C. Selective Hardening for CNNs in FPGAs. *IEEE Trans. Nucl. Sci.* **2019**, *66*, 216–222. [CrossRef]
11. Blower, S.; Rech, P.; Cazzaniga, C.; Kastriotou, M.; Frost, C.D. Evaluating and Mitigating Neutrons Effects on COTS EdgeAI Accel-erators. *IEEE Trans. Nucl. Sci.* **2021**, *68*, 1719–1726. [CrossRef]
12. Kchaou, A.; EL Hadj Youssef, W.; Tourki, R. SEUs Mitigation on Program Counter of the LEON3 Soft Processor. *Int. J. Adv. Comput. Sci. Appl.* **2017**, *8*, 080737. [CrossRef]
13. Saleh, A.; Serrano, J.; Patel, J. Reliability of scrubbing recovery-techniques for memory systems. *IEEE Trans. Reliab.* **1990**, *39*, 114–122. [CrossRef]
14. Sabena, D.; Reorda, M.S.; Sterpone, L.; Rech, P.; Carro, L. On the evaluation of soft-errors detection techniques for GPGPUs. In Proceedings of the 2013 8th IEEE Design and Test Symposium, Marrakesh, Morocco, 16–18 December 2013; pp. 1–6.
15. Oliveira, D.A.G.; Rech, P.; Quinn, H.M.; Fairbanks, T.D.; Monroe, L.; Michalak, S.E.; Anderson-Cook, C.; Navaux, P.O.A.; Carro, L. Modern GPUs radiation sensitivity evaluation and mitigation through duplication with comparison. *IEEE Trans. Nucl. Sci.* **2014**, *61*, 3115–3122. [CrossRef]
16. Kwon, S.; Lee, K.; Kim, Y.; Kim, K.; Lee, C.; Ro, W.W. Measuring error-tolerance in SRAM architecture on hardware accelerated CNN. In Proceedings of the 2016 IEEE International Conference on Consumer Electronics-Asia (ICCE-Asia), Seoul, Republic of Korea, 26–28 October 2016; pp. 1–4.
17. Libano, F.; Wilson, B.; Wirthlin, M.; Rech, P.; Brunhaver, J. Understanding the Impact of Quantization, Accuracy, and Radiation on the Reliability of Convolutional CNNs on FPGAs. *IEEE Trans. Nucl. Sci.* **2020**, *67*, 1478–1484. [CrossRef]
18. Wang, H.B.; Wang, Y.S.; Xiao, J.H.; Wang, S.L.; Liang, T.J. Impact of Single-Event Upsets on Convolutional CNNs in Xilinx Zynq FPGAs. *IEEE Trans. Nucl. Sci.* **2021**, *68*, 394–401. [CrossRef]
19. Arechiga, A.P.; Michaels, A.J. The Robustness of Modern Deep Learning Architectures against Single Event Upset Errors. In Proceedings of the 2018 IEEE High Performance Extreme Computing Conference (HPEC), Waltham, MA, USA, 25–27 September 2018; pp. 1–6.
20. Rech, P.; Aguiar, C.; Frost, C.; Carro, L. An efficient and experimen tally tuned software-based hardening strategy for matrix mul-tiplication on gpus. *IEEE Trans. Nucl. Sci.* **2013**, *60*, 2797–2804. [CrossRef]
21. Ruospo, A.; Gavarini, G.; de Sio, C.; Guerrero, J.; Sterpone, L.; Reorda, M.S.; Sanchez, E.; Mariani, R.; Aribido, J.; Athavale, J. Assessing Convolutional Neural Networks Reliability through Statistical Fault Injections. In Proceedings of the 2023 Design, Automation & Test in Europe Conference & Exhibition (DATE), Antwerp, Belgium, 17–19 April 2023; pp. 1–6.
22. Huang, Z.; Shao, W.; Wang, X.; Lin, L.; Luo, P. Rethinking the pruning criteria for convolutional neural network. *Adv. Neural Inf. Process. Syst.* **2021**, *34*, 16305–16318.

Disclaimer/Publisher's Note: The statements, opinions and data contained in all publications are solely those of the individual author(s) and contributor(s) and not of MDPI and/or the editor(s). MDPI and/or the editor(s) disclaim responsibility for any injury to people or property resulting from any ideas, methods, instructions or products referred to in the content.

electronics

MDPI

Article

Study on Radiation Damage of Silicon Solar Cell Electrical Parameters by Nanosecond Pulse Laser

Sai Li , Longcheng Huang, Jifei Ye *, Yanji Hong *, Ying Wang, Heyan Gao and Qianqian Cui

State Key Laboratory of Laser Propulsion & Application, Space Engineering University, Beijing 101400, China; hebtulisai@163.com (S.L.); huanglc1983@163.com (L.H.); yingwang8971@163.com (Y.W.); gaohy_s@163.com (H.G.); cui_qqq@163.com (Q.C.)

* Correspondence: yjf1981@163.com (J.Y.); hongyanji@vip.sins.com (Y.H.)

Abstract: This experimental study investigates the damage effects of nanosecond pulse laser irradiation on silicon solar cells. It encompasses the analysis of transient pulse signal waveform characteristics at the cells' output and changes in electrical parameters, such as I–V curves before and after laser irradiation under varying laser fluence and background light intensities, and explores the underlying action mechanisms of laser irradiation. The study reveals that as the laser fluence increases up to 4.0 J/cm^2, the peak value of the transient pulse signal increases by 47.5%, while the pulse width augments by 88.2% compared to the initial transient pulse signal. Furthermore, certain parameters, such as open-circuit voltage, short-circuit current, and peak power obtained, from the measured I–V curve indicate a threshold laser fluence for functional degradation of the solar cell at approximately 1.18 ± 0.42 J/cm^2. Results obtained from laser irradiation under different background light intensities underscore the significant influence of background light on laser irradiation of silicon cells, with the most severe damage occurring in the absence of light. Moreover, findings from laser irradiation at multiple locations on the silicon cell demonstrate a linear decrease in the output voltage of the silicon cell with an increase in the number of irradiation points.

Keywords: solar cell; laser; silicon; radiation; electrical parameters

1. Introduction

A laser beam has significant advantages, such as high directionality and monochromaticity, good coherence, and high brightness. It has been widely used in materials (including metal and non-metal) cutting, surface treatment, perforation, and micro-machining, among other applications. Solar cells are photoelectric conversion devices that directly convert light into electrical energy without an external power source. As a photosensitive device, solar cells are susceptible to damage from laser irradiation, resulting in reduced photoelectric conversion efficiency, structural damage, and functional total or partial loss. Increasing the power of incident light on solar cells enhances the output, but high-power monochromatic light irradiation may cause severe damage to them. Currently, the most widely used silicon solar cells present a high absorption rate and a low reflectivity for a 1.1 μm laser, but their photoelectric conversion efficiency is generally between 10% and 20%. Therefore, a significant part of the incident laser beam energy converts into thermal energy that heats and damages the solar cell. When the incident laser beam energy is much greater than the energy that the solar cell can absorb, the incident laser beam will increase the temperature of the cell, leading to a significant decrease in efficiency. The research on laser irradiation of solar cells began with the concept of laser wireless power transmission. Glaser proposed the idea of establishing a solar power collection system in space to transmit abundant solar energy to ground-based solar cells through laser transmission. The American Explorer detector discovered that high-energy charged particle radiation in space can cause damage to the performance of solar cells. With the further development and application of laser technology, attention has gradually focused on researching the damage and protection of laser-irradiated solar cells.

In 1973, Matsuoka and Usami first conducted a series of studies on the laser irradiation of solar cells. The experiments found that excessive power of pulsed lasers leads to a decrease in the peak power of solar cells [1–3]. Recently, research has been conducted on the irradiation damage of solar cells using continuous laser [4–9] and pulsed laser [10–12], accounting for the effects of different wavelengths [13–15], employing different pulse widths [16,17], and varying the laser parameters, such as laser fluence and frequency [18–22]. The damage effect of laser irradiation on silicon materials used in solar cells and its mechanism have been the focus of research [23–25]. The investigation of the damage to silicon materials mainly focuses on the damage caused by heat and stress. However, this does not directly reflect the electrical damage of silicon-based electronic devices. Most of the research on laser-irradiated solar cells treats the damage threshold, temperature response, stress damage, and damage morphology of solar cells under different laser parameters or irradiation modes. Photosensitive devices, such as silicon-based photodiodes [26,27], present significant transient electrical signals under laser irradiation. These transient signals are the direct response of the device to the laser, and their characteristics can indicate the interaction process between the laser and the device. Analogously, solar cells can generate photoelectric signals even under weak laser conditions. These signals are crucial for analyzing the intrinsic mechanism of weak laser-induced damage to solar cells. However, the characteristics of these laser-induced transient pulse signals have not received much attention.

In this study, we developed an acquisition circuit to detect the transient pulse signal resulting from laser irradiation on silicon cells. We conducted experiments to investigate the characteristics of the transient pulse signal under different laser fluences, varying background light intensities, and multi-point irradiation scenarios on the same cell. Simultaneously, we obtained the I–V curve of the cell under different experimental conditions and analyzed the underlying mechanism of laser action. The photoelectric signal extraction method employed in this research is innovative, and the findings can offer valuable insights for implementing damage protection measures for solar cells to resistant lasers.

2. Device and Methods

2.1. Device and Signal Acquisition Circuit

The operational mechanism of solar cells is based on the photovoltaic effect of semiconductor materials. This effect describes the generation of charge carriers within semiconductor materials when exposed to light irradiation, and the charge carriers, upon separation, accumulate at the respective ends of a PN junction, thus establishing a potential difference. When an external circuit is connected to both ends of the PN junction, the photocurrent flows through the load. Figure 1 illustrates the operational concept of a silicon solar cell. When the energy of incident photons (hv) surpasses the bandgap width (E_g) of the semiconductor material, numerous electron–hole pairs get excited. If these pairs form within the vicinity of the PN junction's diffusion length, they can potentially diffuse into the depletion region of the PN junction before undergoing recombination. Under the strong electric field in the depletion region, the electrons migrate toward the N region while holes move toward the P region. This migration leads to a negative charge in the N region and a positive charge in the P region. The electrodes at the sides of the PN junction connect to the load, and the photocurrent flows through the circuit, thus accomplishing power output.

The experimental setup utilized a polycrystalline silicon solar cell measuring 45 mm × 45 mm. The silicon cell exhibits operating voltage, open-circuit voltage, and short-circuit current of 1 V, 1.2 V, and 0.2 A, respectively, with a conversion efficiency of 18%.

Figure 2 depicts the acquisition circuit diagram for the output voltage signal at the load end of the solar cell. R_0 is a small resistance, while R_m denotes the optimal matching resistance corresponding to the cell's highest output power. The resistance calculated according to the measured I–V curve is used as a reference for R_m. For the silicon cell employed in this experiment, the circuit features R_0 and R_m values of 0.1 ohms and 4.3 ohms, respectively.

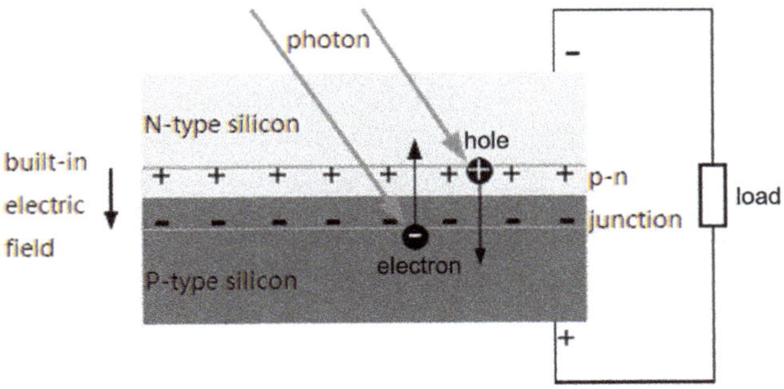

Figure 1. The working principle of a silicon solar cell.

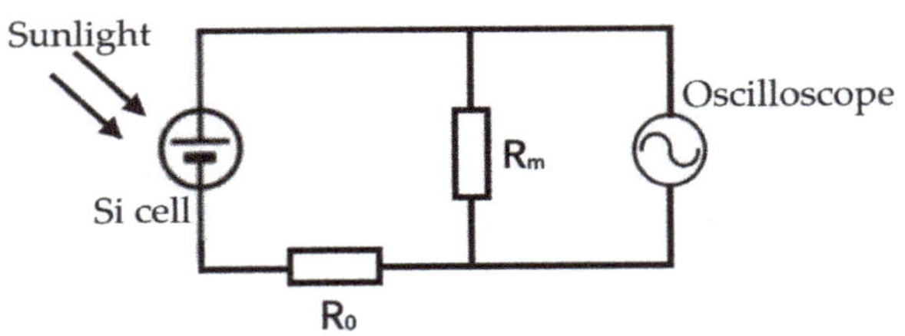

Figure 2. The signal acquisition circuit diagram of the Si cell.

2.2. Methods

The laser employed was a neodymium yttrium aluminum garnet (Nd: YGA) laser with a wavelength of 1064 nm and a pulse width of 10 ns. Figure 3 shows the schematic diagram of the experimental setup. The laser beam section is first shaped by the aperture and then sequentially passes through the attenuator and spectroscope before finally reaching the focusing lens. After being focused by a lens, the laser incidences on the surface of the silicon cell, with a section of $\Phi = 4$ mm. The attenuator adjusts the laser beam intensity, and the spectroscope divides the beam into two paths: one for irradiating the solar cell and the other for laser beam intensity measurement. At the sample location, a Solar-500 xenon lamp (emitting spectra in the range of 200 nm–2500 nm) is used as a solar simulator to provide background light when the Si cell is working. The light intensity of the lamp is adjustable by controlling the current. Throughout the experiment, the voltage probe is connected across the sampling resistor to gather the output voltage of the Si cell and the transient pulse signal produced upon laser irradiation.

This experimental strategy involved conducting laser irradiation experiments with varying laser beam energies, background light intensities, and laser irradiation positions. The experiment involved measuring the voltage–current characteristic curves of the cell both before and after laser irradiation, alongside capturing transient voltage signals generated during the irradiation process. We measured the voltage–current characteristic curve of the cell using a Keithley 2450 instrument. The voltage signals were recorded by the oscilloscope. Electrical parameters of the cell, such as open-circuit voltage, short-circuit current, and peak power, were extracted from the volt–ampere characteristic curve. Subsequently, the analysis focused on the changes in these parameters following laser irradiation.

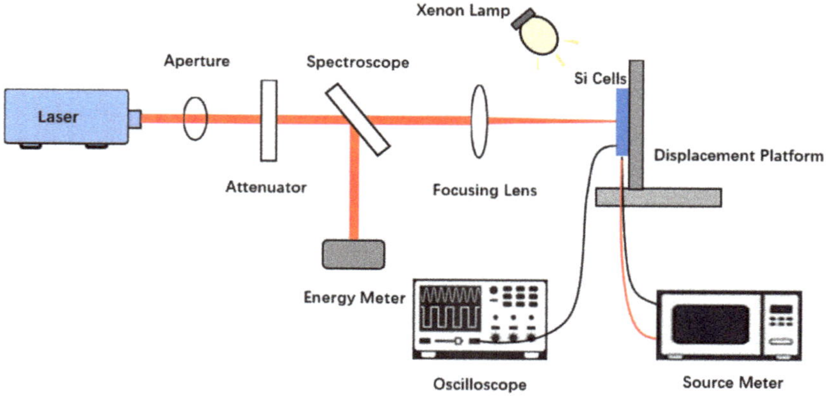

Figure 3. Schematic diagram of the experimental setup.

3. Results

3.1. Results at Different Laser Fluences

Figure 4 shows the transient pulse signals generated at the silicon cell output when irradiated by a single-pulse laser; each curve corresponds to a different laser fluence. The curves show that when the laser fluence is 0.24 J/cm^2, the laser does not cause any disturbance to the cell. When the laser fluence is 0.76 J/cm^2 or more, it generates a disturbance voltage signal at the output of the Si cell. After the disturbance, the voltage at the output of the cell returns to the level before irradiation. Upon reaching or surpassing 1.6 J/cm^2, the disturbance signal increases in duration and peak voltage. Furthermore, following laser irradiation, the cell output voltage fails to recover its pre-irradiation level, suggesting a notable decline in cell performance. We conclude that the interference threshold of laser-irradiated solar cells is 0.76 $\pm$ 0.26 J/cm^2, and the damage threshold is about 1.18 $\pm$ 0.42 J/cm^2.

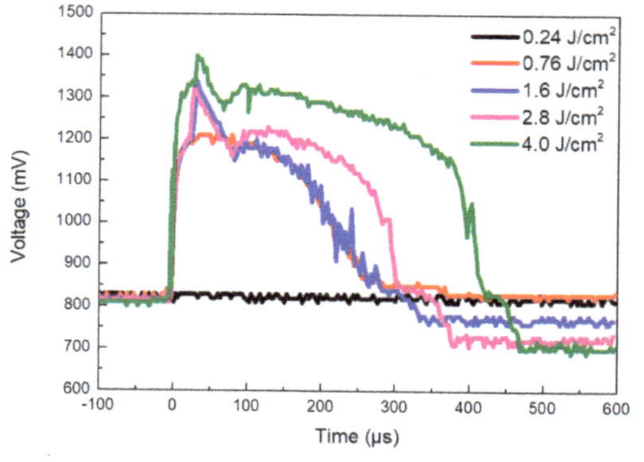

Figure 4. Transient pulse signals generated by Si cell at different laser fluences.

Figure 5a,b show the I–V and P–V curves of the Si cell before and after single-pulse laser irradiation at different laser fluences. Table 1 presents the output voltage of the Si cell, the open-circuit voltage, the short-circuit current, and the peak power extracted from Figure 5. Table 1 indicates that when the laser fluence is less than or equal to 0.76 J/cm^2, it does not cause significant damage to the cell's performance. Increasing the laser fluence from 1.6 J/cm^2, the output voltage and peak power of the Si cell show a significant decrease, the open-circuit voltage also decreases, and the short-circuit current does not change considerably.

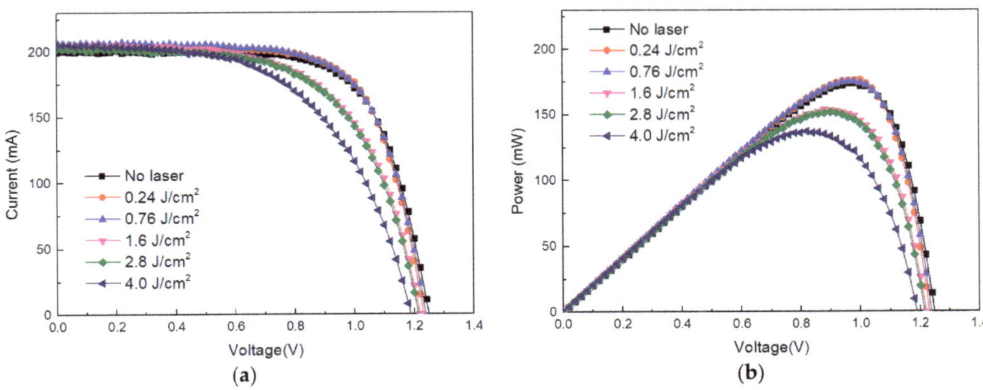

Figure 5. Electrical parameters of Si cells at different laser fluences: (**a**) I–V curve; (**b**) P–V curve.

Table 1. Electrical performance parameters of the Si cell under laser irradiation.

Laser Fluence ($J\cdot cm^{-2}$)	Output Voltage (mV)	Open-Circuit Voltage (V)	Short-Circuit Current (mA)	Peak Power (mW)
0	820	1.24	200	174
0.24	820	1.24	204	176
0.76	820	1.24	207	175
1.6	780	1.22	203	153 (−12.1%)
2.8	740	1.22	202	152 (−12.6%)
4.0	720	1.18	205	137 (−21.3%)

3.2. Results at Different Background Light Intensities

Figure 6 shows the transient pulse signal generated at the output of the Si cell when irradiated with a single-pulse laser at different background light intensities. The figure indicates that regardless of the background light (corresponding to different initial voltages of the cell), the laser irradiation will cause a larger transient pulse signal at the cell's output. When there is background light, the amplitudes of the transient pulse signal generated by laser irradiation under 550 W/cm^2 and 1100 W/cm^2 are almost the same (about 1400 mV). When there is no background light, the amplitude of the transient pulse signal is significantly higher (approximately 1540 mV). Additionally, with the increase in background light intensity, the pulse width of the transient pulse signal gradually widens.

Figure 7a,b show the I–V and P–V curves of the Si cell under single-pulse laser irradiation at different background light intensities. Table 2 indicates the output voltage of the Si cell under laser irradiation, as well as the open-circuit voltage, short-circuit current, and peak power, as extracted from Figure 7. Table 2 indicates that background light is an important influencing factor on the characteristics of silicon cells under laser irradiation. Moreover, the performance of the cells is more severely damaged when there is no light. Coinciding with this observation, it is evident that as the light intensity decreases, the cell's peak power and open-circuit voltage decline, while the short-circuit current initially remains constant before eventually increasing.

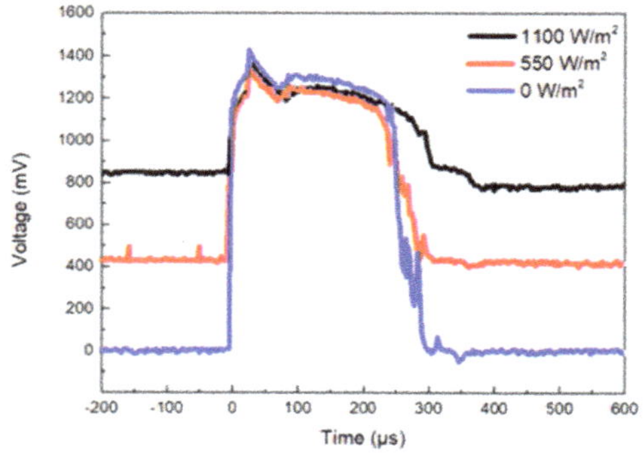

Figure 6. Transient pulse signals generated by the Si cell at different background light intensities.

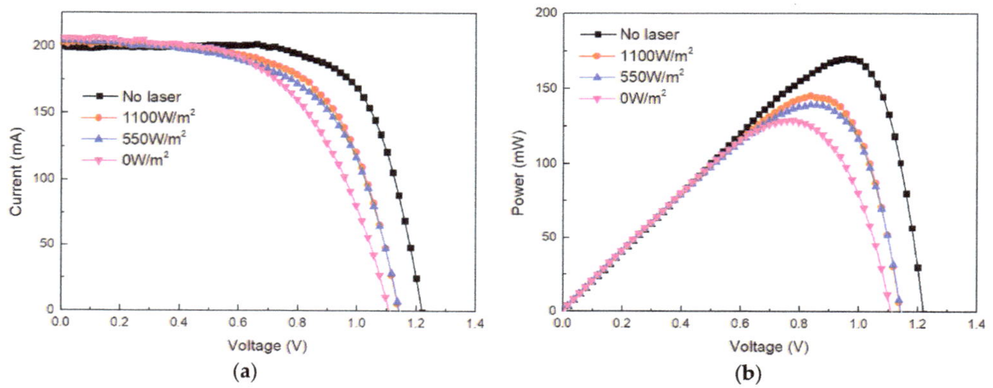

Figure 7. Electrical parameters of Si cells at different background light intensities: (**a**) I–V curve; (**b**) P–V curve.

Table 2. Electrical performance parameters of the Si cell under laser irradiation.

Background Light Intensity (W/m^2)	Open-Circuit Voltage (V)	Short-Circuit Current (mA)	Peak Power (mW)
1100 (No laser)	1.22	200	171
1100	1.14	200	145 (−15.2%)
550	1.14	200	140 (−18.1%)
0	1.12	202	129 (−24.6%)

3.3. Results of Multi-Point Irradiation

Figure 8 shows the ablation points generated by laser irradiation on Si cells during the experiments. A total of 6 times of laser irradiation were carried out, and the order of ablation points is shown in the figure.

Figure 9a shows the transient pulse signal generated at the output of the Si cell at different irradiation positions. The figure indicates that each transient pulse signal is similar. The peak values of these transient pulse signals are 1350 mV, 1390 mV, 1370 mV, 1410 mV, 1360 mV, and 1410 mV, successively. Figure 9b depicts the output voltage of the Si cell after each laser irradiation. The Si cell output voltage decreases linearly with an increase in the number of laser irradiation points. After applying six pulse lasers, the Si cell output voltage dropped from 855 mV to 455 mV, which means that the performance of the Si cell decreased by about 47%.

Figure 8. The ablation points generated by laser irradiation on Si cell.

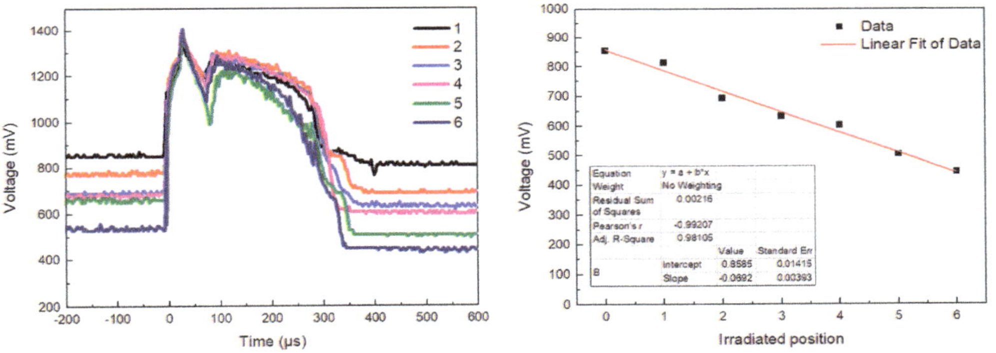

(**a**) (**b**)

Figure 9. (**a**) Transient pulse signals generated by the Si cell at multiple different irradiation positions; (**b**) output voltage of the Si cell after each laser irradiation.

Figure 10a,b depict the I–V and P–V curves of Si cells following laser irradiation at various positions. These figures demonstrate that each laser irradiation induces a degree of performance degradation in the cell.

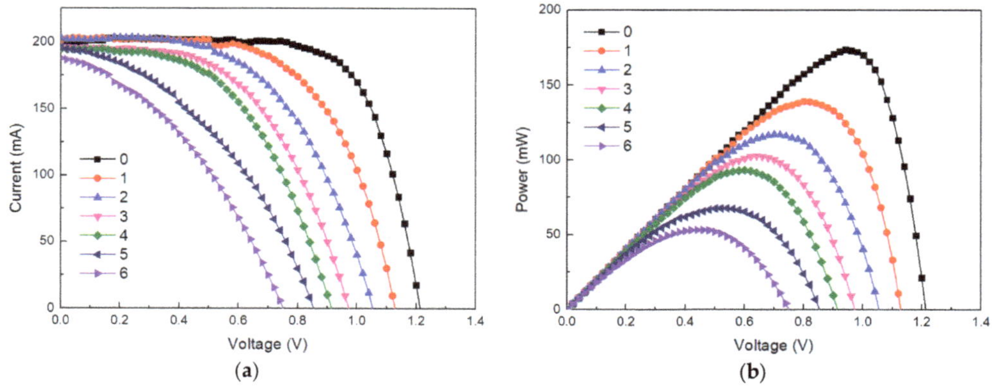

Figure 10. Electrical parameters of Si cell after laser irradiation at different positions: (**a**) I–V curve; (**b**) P–V curve.

4. Discussions

4.1. The Influence of Laser Fluence

Figure 11 illustrates the variation in peak voltage and pulse width of the transient voltage signal at the output of the Si cell with increasing laser fluence. The graph reveals three distinct stages in the changes of these parameters:

- Initially, upon increasing the laser fluence, the peak value of the transient pulse signal rises by 32.5% compared to the initial pulse signal, while the pulse width remains nearly unchanged.
- Subsequently, although the peak value of the transient pulse signal remains relatively constant, its pulse width expands by 29.4% compared to the initial signal.
- Finally, with further increases in laser fluence, the peak value and pulse width of the transient pulse signal increase by 47.5% and 88.2%, respectively, compared to the initial signal.

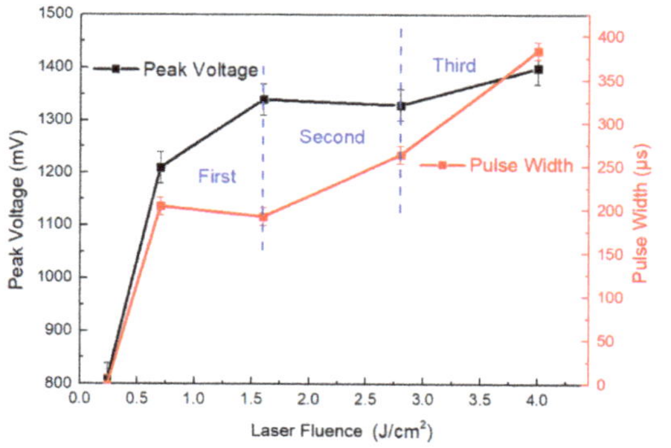

Figure 11. Transient voltage signal characteristic parameters: peak voltage (left), pulse width (right).

The wavelength of light absorbed by Si solar cells typically falls between 300 nm and 1200 nm. Given that the photon energy of the 1064 nm laser employed in the experiment exceeds the bandgap width of silicon, the laser radiation induces the photoelectric effect within the cell. This process generates a significant number of charge carriers, which are subsequently collected and produce transient pulse signals at the cell's output. Figure 11 illustrates that as the laser fluence rises, the amplitude of the transient disturbance signal at the cell's output increases. Moreover, as the laser fluence continues to escalate, the amplitude gradually approaches saturation. When subjected to light, Si solar cells generate an internal electric field. Charge carriers produced by laser irradiation are collected by drift motion under the action of the internal electric field. The amplitude and collection time of the transient pulse signal are determined by carrier concentration and drift velocity.

In the initial stage, the carrier concentration rises with the laser fluence, yet the charge collection rate does not reach its maximum. As a result, the amplitude of the transient disturbance signal increases while the pulse width remains relatively constant. Subsequently, as the laser fluence continues to increase, a significant number of photogenerated carriers are produced within the cell. At this point, the carrier collection rate saturates, leading to a consistent amplitude of the disturbance signal. Consequently, the collection time extends, resulting in an expansion of the pulse width. In the final stage, with further increases in laser fluence, more charge carriers are generated within the cell. This results in enhanced charge collection, which, combined with the effects of drift and diffusion, further prolongs the collection time.

4.2. The Influence of Background Light Intensity

The generation of carriers within the cell is contingent upon the laser fluence, whereas background light intensity primarily influences the collection process of carriers. The number of photogenerated carriers in the cell is proportional to the laser fluence. When there is background light, the silicon cell absorbs the light energy and converts it into electricity. The current generated by the light forms a loop through the load resistance. At this time, a small portion of the carriers induced by the laser are recombined through the current path, while more carriers are collected at the resistor. Without background light, the sampling resistance collects almost all of the charge carriers. Therefore, in this case, the amplitude of the transient pulse signal generated by laser irradiation is larger. Table 2 indicates that the cell's peak power and open-circuit voltage decrease as the background light diminishes. This exhibits the dramatic decrease in the Si cell performance with background light diminution.

Figure 12 shows the ablation morphology of the Si cell by a single-pulse laser under different background light intensities. As the background light intensity decreases, the cell damage becomes more severe. This is because when there is background light, the recombination of charge carriers through the pathway can release some heat, so the ablation is not as severe. Laser irradiation can cause ablation damage inside the cell, leading to internal leakage current, which reduces the parallel resistance of the cell and causes a decrease in the open-circuit voltage of the solar cell. The series resistance does not affect the open-circuit voltage of solar cells, but as the resistance value increases, it will cause a voltage drop between the terminals. When it exceeds a critical value, it will lead to a decrease in the short-circuit current.

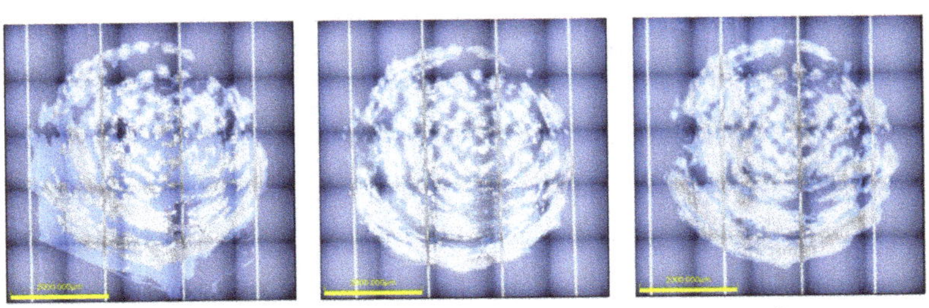

1100W/m^2 550W/m^2 0

Figure 12. Ablation morphology of laser-irradiated cells under different background light intensities.

4.3. The Influence of Multi-Point Irradiation

The variations in certain parameters, such as open-circuit voltage, short-circuit current, and peak power of the cell, under multiple pulse laser irradiations are shown in

Figure 13a–c, respectively. The figures show that as the number of irradiated lasers increases, the open-circuit voltage of the cell decreases approximately linearly; the short-circuit current of the cell initially remains almost unchanged and then slightly decreases, and the peak power significantly decreases with a turning point in the rate of decrease. The point at which the short-circuit current begins to decrease corresponds to the inflection point of the peak power decrease rate.

In a state of constant illumination, the photocurrent of a solar cell remains unaffected by its operating condition; thus, it is akin to a constant current source. Additionally, the p–n junction in the cell can be regarded as an ideal diode. Therefore, a typical solar cell circuit is equivalent to a parallel circuit of a current source and diode. The material's intrinsic resistance dictates the presence of parasitic resistances, comprising the parallel resistance R_{sh} and the series resistance R_s. The external electrical device is represented as a load resistance R_L. Figure 14 depicts the equivalent circuit of a solar cell.

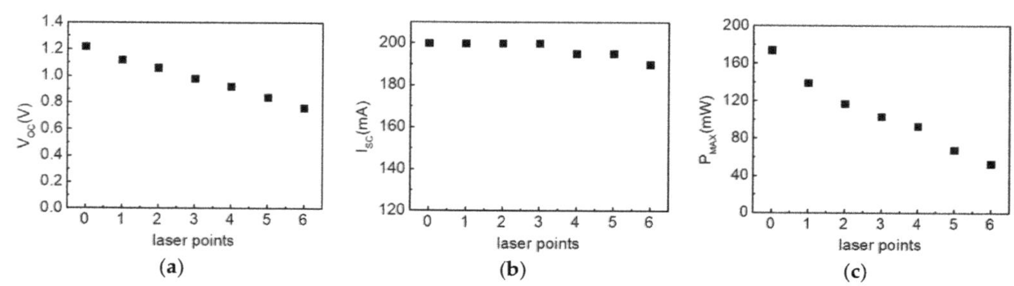

Figure 13. Characteristics of Si cell's electrical parameters under laser irradiation at different positions: (**a**) open-circuit voltage; (**b**) short-circuit current; (**c**) peak power.

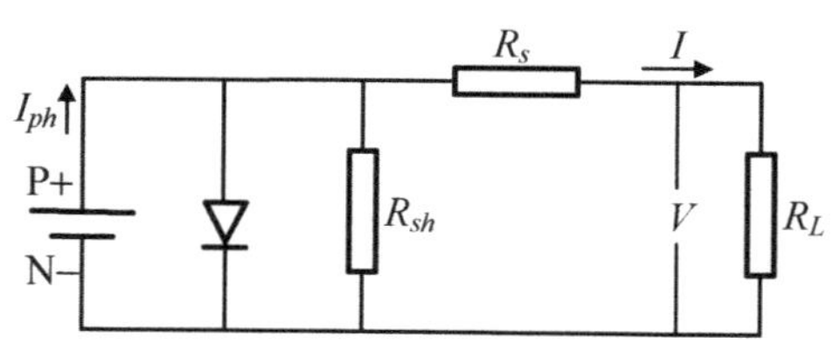

Figure 14. The equivalent circuit of a solar cell.

Under constant external conditions, the performance of a cell is predominantly affected by the parallel R_{sh} and series R_s resistances. We estimated the parallel resistance of the cell by calculating the slope of the I–V characteristic curve near the short-circuit current [28]. Analogously, we estimated the series resistance by calculating the slope of the I–V characteristic curve of the open-circuit voltage [29]. Figure 15a,b show the variation patterns of Si cell parallel resistance and series resistance obtained through theoretical calculations, respectively [30].

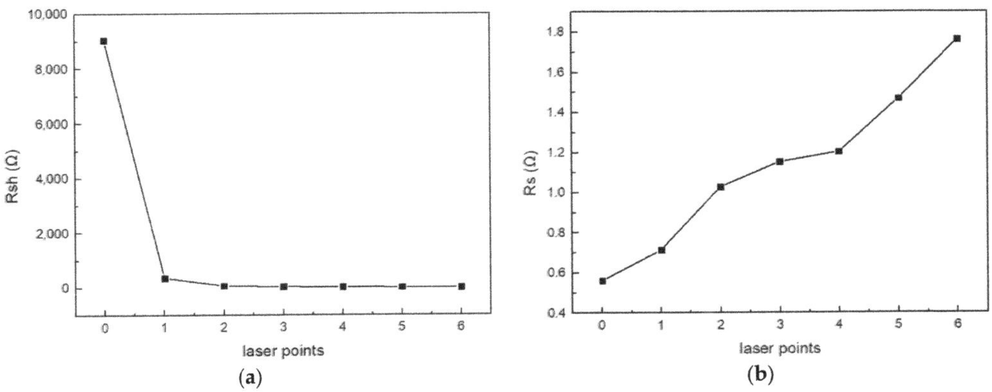

Figure 15. Parallel and series resistance of the Si cell under laser irradiation at different positions: (**a**) parallel resistance; (**b**) series resistance.

The parallel resistance R_{sh} in a cell is typically due to manufacturing defects, such as PN junction leakage current. This leakage current includes electrodes that bypass the edge of the cell and internal leakage current caused by crystal defects and foreign impurities in the junction area. The resistance value is usually several thousand ohms. The decrease in the parallel resistance will reduce the filling factor of the solar cell because the parallel resistance will affect the maximum power of the solar cell and make it deviate from the ideal maximum power value. Meanwhile, the reduction of parallel resistance will decrease the open-circuit voltage of the solar cell. However, it does not affect the short-circuit current. From Figure 15a, we can observe that each laser irradiation causes the parallel resistance of the cell to decrease to a certain extent. This is because the cell undergoes ablation damage and leakage under laser irradiation, resulting in a decrease in parallel resistance and the cell's open-circuit voltage.

The cell's series resistance R_s originates from four main contributions: the inherent resistance of the semiconductor material, the contact resistance between the metal electrode and the semiconductor material, the resistance through the diffusion layer, and the resistance of the metal electrode itself. Among them, the transverse resistance of the diffusion layer is the main form of series resistance, and the series resistance is usually less than 1 Ω. The increase in series resistance will also reduce the filling factor of solar cells because the filling factor is the ratio of the maximum output power of solar cells to the ideal output power, of which the ideal output power is the output power based on the assumption that the solar cell has no internal resistance. In addition, when the resistance value of series resistance is too large, it will also reduce the short-circuit current, but it does not affect the open-circuit voltage.

5. Conclusions

This experimental study investigated the damage effects of nanosecond pulse laser exposure on silicon solar cells under varying laser fluences, background light intensities, and multi-point irradiation. Key findings include:

(1) The interference threshold of laser-irradiated solar cells is 0.76 ± 0.26 J/cm^2, and the damage threshold is about 1.18 ± 0.42 J/cm^2;

(2) In the presence of background light, Si cells can promptly collect the carriers generated by the laser beam and effectively release the heat of the charge carriers during the process. Therefore, the performance of the Si cell is most severely damaged when there is no background light;

(3) Following six consecutive pulse laser irradiations, the cell's output voltage decreases significantly from 855 mV to 455 mV. This reduction in open-circuit voltage and short-

circuit current is attributed to changes in the values of parallel and series resistance in the cell after laser irradiation.

This article analyzed the damage characteristics and internal mechanism of the above three factors on laser-irradiated solar cells based on the transient signal generated during laser induction and the solar cell's I–V curve parameters. The above conclusions can provide theoretical support and protection design ideas for solar cells to resist laser damage.

Author Contributions: Conceptualization, S.L. and J.Y.; methodology, Y.H.; validation, S.L. and L.H.; investigation, Y.W.; resources, Y.H.; data curation, H.G. and Q.C.; writing—original draft preparation, S.L.; writing—review and editing, J.Y.; supervision, Y.H. All authors have read and agreed to the published version of the manuscript.

Funding: This research received no external funding.

Data Availability Statement: Data are contained within the article.

Conflicts of Interest: The authors declare no conflict of interest.

References

1. Matsuoka, Y.; Usami, A. Laser Damage of Silicon Solar Cells. *J. Nucl. Sci. Technol.* **1973**, *10*, 61–62. [CrossRef]
2. Matsuoka, Y. Normal laser damage of silicon solar cells without phase change. *Appl. Phys. Lett.* **1974**, *25*, 574. [CrossRef]
3. Matsuoka, Y.; Usami, A. Laser damage of silicon solar cells with different surface states. *Opto-Electronics* **1974**, *6*, 217–223. [CrossRef]
4. Miyakawa, H.; Tanaka, Y.; Kurokawa, T. Photovoltaic cell characteristics for high intensity laser light. *Fuel Energy Abstr.* **2005**, *86*, 253–267. [CrossRef]
5. Li, G.; Zhang, H.; Zhou, G.; Lu, J.; Zhou, D. Research on influence of parasitic resistance of InGaAs solar cells under continuous wave laser irradiation. *J. Phys. Conf. Ser.* **2017**, *844*, 12–14. [CrossRef]
6. Dou, P.C.; Feng, G.B.; Zhang, J.M.; Song, M.Y.; Zhang, Z.; Li, Y.P.; Shi, Y.B. Laser induced non-monotonic degradation in short-circuit current of triple-junction solar cells. *Opt. Commun.* **2018**, *417*, 37–41. [CrossRef]
7. Zhang, Y.; Li, Y.; Li, Y.; Zhao, G.; Chen, M. Research of silicon solar cells performance after being irradiated by high power laser. In *High-Power Lasers and Applications VIII*; SPIE: Beijing, China, 2016; Volume 10016, pp. 8–15.
8. Li, G.; Zhang, H.; Wang, C.; Lu, J. Effect of 1070 nm laser uniformity on temperature distribution and performance of In0.3Ga0.7As solar panel. In *Fifth International Symposium on Laser Interaction with Matter*; SPIE: Changsha, China, 2019; Volume 11046, pp. 109–117.
9. Guo, W.; Chang, H.; Yu, C.H.; Li, M.Y. Damage characteristics of continuous-wave laser ablation triple-junction solar cells. *J. Laser Appl.* **2022**, *34*, 042038. [CrossRef]
10. Lowe, R.A.; Landis, G.A. Response of photovoltaic cells to pulsed laser illumination. *IEEE Trans. Electron Devices* **1995**, *42*, 744–751. [CrossRef]
11. Qiu, D.; Wang, R.; Cheng, X.; Zhang, Z.; Jiang, T. Damage effect of monocrystalline silicon solar cells under ultrashort pulse laser irradiations. *Infrared Laser Eng.* **2012**, *41*, 112–115.
12. Ke, D. *Study on the Electromechanical Response Characteristics of Solar Arrays Irradiated by Pulsed Laser and Simulated Sunlight*; Shenyang Ligong University: Shenyang, China, 2023.
13. Zhu, R. *Researches on the Irradiation Effects of Single Junction GaAs/Ge and Monocrystalline Silicon Solar Cells under Laser*; National University of Defense Technology: Changsha, China, 2014.
14. Höhn, O.; Walker, A.W.; Bett, A.W.; Helmers, H. Optimal laser wavelength for efficient laser power converter operation over temperature. *Appl. Phys. Lett.* **2016**, *108*, 971–974. [CrossRef]
15. Li, Y.; Zhang, J.; Dou, P.; Shi, Y.; Feng, G. Thermal damage mechanism of single junction GaAs solar cells irradiated by continuous wave laser. *Infrared Laser Eng.* **2018**, *47*, 97–102.
16. Wang, X.; Shen, Z.H.; Lu, J.; Ni, X.W. Laser-induced damage threshold of silicon in millisecond, nanosecond, and picosecond regimes. *J. Appl. Phys.* **2010**, *108*, 033103. [CrossRef]
17. Zhu, R.; Wang, R.; Cheng, X.; Xu, Z. Research of concentric iridescent ring around the laser-induced pits on the solar cell surface. *Appl. Surf. Sci.* **2015**, *336*, 375–379. [CrossRef]
18. Tian, X.; Xiao, S.; Tao, S.; Yuan, Z.; Zhou, Y. Damage threshold research of monocrystalline silicon solar cells under femtosecond laser illumination. *Infrared Laser Eng.* **2014**, *43*, 676–680.
19. Wu, W.; Chen, R.; Hua, Y.; Liu, H.; Xue, Q. The damage characteristics study on the solar cell by femtosecond laser. *J. Funct. Mater.* **2015**, *46*, 17049–17057.
20. Xue, Q.; Wu, W.; Ye, Y.; Liu, H.; Chen, R.; Hua, Y. Property Degradation of GaAs/Ge Solar Cells after Femtosecond Laser Irradiation. *Laser Optoelectron. Prog.* **2015**, *52*, 122–128.
21. Chang, H.; Chen, Y.; Guo, W. Damage characteristics of the solar cells irradiated by nanosecond pulsed lasers and the effects on photoelectric conversion. *Infrared Laser Eng.* **2021**, *50*, 20210296-1.

22. Yang, Y.; Lyu, T.; Zhang, R. Effect of repetition frequency laser on the response characteristics of silicon solar cell. *Laser J.* **2024**, *1*, 70–74.
23. Guo, M.; Zhang, Y.; Zhang, W.; Li, H. Effect of spot size on laser damage single crystal silicon. In *Conference on Applied Optics and Photonics China*; SPIE: Beijing, China, 2019; p. 11333.
24. Jia, Z.; Wang, W.; Li, X.; Hao, L. Millisecond laser-induced damage process of (001) silicon wafer. *Opt. Eng.* **2021**, *60*, 097101. [CrossRef]
25. Ohmura, E.; Fukumitsu, K.; Kumagai, M.; Morita, H. Analysis of Internal Modified-Layer Formation Mechanism into Silicon Single Crystal with Nanosecond Laser. *Comput. Mater. Sci.* **2019**, *160*, 420–429. [CrossRef]
26. Gao, L.; Wei, Z.; Wang, D.; Ma, J.; Liang, C. Experimental study of 1064nm CW laser ablation biased silicon-based PIN photodiodes. In *Advanced Laser Technology and Applications*; SPIE: Beijing, China, 2021; p. 12060.
27. Wang, T.; Li, P.; Yu, X.; Wang, K.; Wang, D.; Zhang, Y.; Li, C. High-energy hundred-picosecond fiber-solid hybrid laser and its application in laser-induced damage in PIN photodiode. *Laser Phys.* **2020**, *30*, 036004. [CrossRef]
28. Chen, T.; Tu, J.; Wang, D. A method for measuring parallel resistance of solar cells. *Semicond. Optoelectron.* **1998**, *19*, 3.
29. Liu, Q.; Guo, K.; Mao, M. A photovoltaic fault detection method based on series equivalent resistance. *J. Sol. Energy* **2020**, *41*, 8.
30. Huang, L.; Ye, J.; Li, S.; Li, L.; Wang, D. Effect of multiple pulsed laser irradiation on resistance of silicon cells. In Proceedings of the Sixth International Symposium on Laser Interaction with Matter 124590D, Ningbo, China, 22 December 2022.

Disclaimer/Publisher's Note: The statements, opinions and data contained in all publications are solely those of the individual author(s) and contributor(s) and not of MDPI and/or the editor(s). MDPI and/or the editor(s) disclaim responsibility for any injury to people or property resulting from any ideas, methods, instructions or products referred to in the content.

electronics

MDPI

Article

Model Parameters and Degradation Mechanism Analysis of Indium Phosphide Hetero-Junction Bipolar Transistors Exposed to Proton Irradiation

Xiaohong Zhao [1,*], Hongwei Wang [1,*], Yihao Zhang [1], You Chen [1], Siyi Cheng [1], Xing Wang [1], Fang Peng [1], Yongjian Yang [1], Guannan Tang [2], Yurong Bai [3] and Shaowei Sun [4]

1 School of Aviation Engineering, Air Force Engineering University, Xi'an 710051, China
2 School of Air Defense and Antimissile, Air Force Engineering University, Xi'an 710051, China
3 School of Nuclear Science and Technology, Xi'an Jiaotong University, Xi'an 710049, China
4 LONGi Green Energy Technology Co., Ltd., Xi'an 710018, China
* Correspondence: zhaoxiaohong0608@163.com (X.Z.); hww0818@163.com (H.W.)

Abstract: The degradation properties of Indium phosphide hetero-junction bipolar transistors (InP HBTs) under proton irradiation are studied and modelled using a compact model for pre-irradiation, post-irradiation, and post-annealing. The variation rates of the model parameters, such as the base–emitter saturation current (I_{SE}) and ideality factor in the ideal region (N_E) in the forward Gummel characteristics, the zero-biased capacitance (C_{je}) and the grading factor (M_{jer}) in the BE junction capacitance, and the transit time parameter in the base region (T_{fb}), are analysed to delve into the degradation mechanism induced by proton irradiation. The displacement damage, induced by proton irradiation in the space charge region of the base–emitter junction and in the quasi-neutral bulk base region, is found to be responsible for the decrease in current gain and cut-off frequency. After annealing, the variation rates of the parameters decrease significantly compared to post-irradiation. This suggests that the recombination of unstable defects leads to a slight recovery in the degradation characteristics of InP HBTs after a period of annealing.

Keywords: hetero-junction bipolar transistor; degradation mechanism; model simulation; irradiation defects

1. Introduction

Indium phosphide hetero-junction bipolar transistors (InP HBTs) are extensively used in aerospace systems, military communications and satellites owing to their exceptional material properties and high-frequency characteristics within various millimetre-wave devices [1–3]. However, in the harsh irradiation environment of space, InP HBTs are susceptible to single-event transients (SETs) triggered by heavy ions [4–6] and displacement damage induced by high-flux protons [7,8]. While our previous work has covered research on the single-event effect [9], this paper will concentrate on displacement damage.

Multiple energetic protons can induce displacement damage when striking micro-electronic devices, particularly in aerospace systems operating within the inner Van Allen radiation belts over prolonged periods of time [10]. These displacement defects include host atoms displacement, lattice damage and the formation of deep-level traps [11,12]. The reported studies primarily delineated and analysed experimental phenomena [7,13,14], with limited literature delving into the irradiation-induced degradation mechanism through compact models. Lawal O M et al. discussed the Simulation Program with Integrated Circuit Emphasis (SPICE) model parameters of a pre- and post-irradiated silicon bipolar junction transistor (Si BJT) [15]. Petrosyants K et al. explored the SPICE model of a Si BJTs and silicon germanium (SiGe) hetero-junction bipolar transistor (HBTs), taking into account the total irradiation dose effects with gamma-ray, proton, neutron, and electron irradiation [16–18]. Van Uffelen M et al. studied the direct current (DC) characteristics of SiGe HBTs at gamma doses of up to 4 MGy using the SPICE model [19]. Zhang J C et al. investigated the impact of gamma irradiation on the DC and alternating current (AC) characteristics of gallium arsenide (GaAs) and InP hetero-junction bipolar transistors with the VBIC and Keysight models [20–22]. Thus, the relevant literature mainly focuses on model parameter analysis of gamma irradiation for Si BJTs, GaAs HBTs, SiGe HBTs and InP HBTs, while lacking the model parameter analysis of proton irradiation. Although the analysis method is similar, it is crucial that proton irradiation and gamma irradiation interact differently with semiconductor materials, resulting in distinct damage mechanisms and affecting the various model parameters. The effects of Gamma irradiation on devices cannot be used as an equivalent analysis for the effects of proton irradiation on InP HBT. The degradation characteristics reflected by the model parameters are different. Furthermore, based on the model parameters, the analysis of the influence of proton irradiation on the circuit are also different. Therefore, it is essential to study the compact model analysis of proton irradiation for InP HBTs.

Moreover, the compact model serves as a vital bridge between the device and circuit design. It not only helps to predict the radiation characteristics of a device, but also provides a potent tool for designing anti-radiation circuits. Therefore, in this paper, the influence of proton-induced displacement damage on the model parameters is studied to explore the degradation mechanisms for pre-irradiation, post-irradiation and post-annealing. The paper consists of four sections. The description of experiments is presented in Section 2. The introduction of the model and the analysis of degradation mechanism are discussed in Section 3. The conclusions are presented in Section 4.

2. Experiment Details

The devices under test (DUT) are provided by the Institute of Microelectronics in the Chinese Academy of Sciences. Figure 1a shows the detailed doping information and vertical layer thickness of the NPN-type InP HBT. Figure 1b shows the schematic diagram of the triple mesa process and bisbenzocyclobutene (BCB) planarisation passivation structure. For the NPN-type InP HBTs, the non-alloyed Ti/Pt/Au (15/15/400 nm) was used as an n-type Ohmic contact, and Pt/Ti/Pt/Au (2/15/15/120 nm) served as p-type Ohmic contact [23]. Additionally, a 2 μm layer of Au was deposited on the electrodes to act as pads. The emitter area of InP HBT is $1 \times 15\ \mu m^2$.

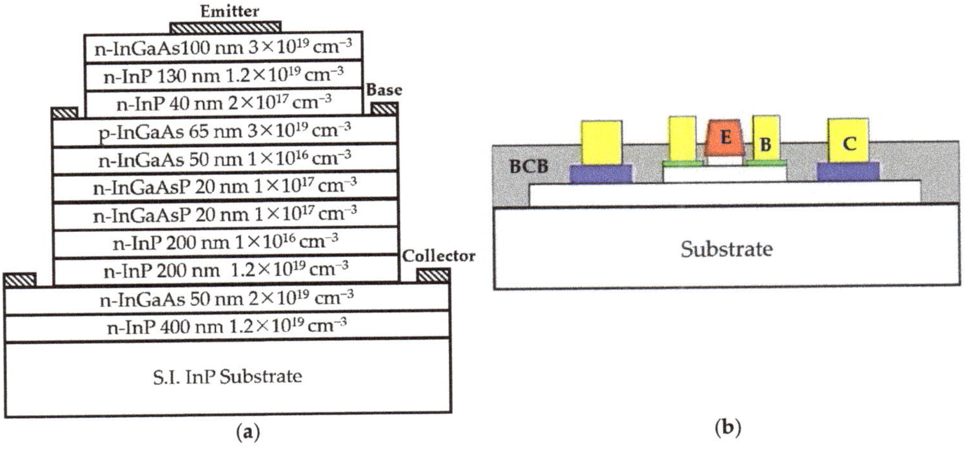

Figure 1. The structure diagram of InP/InGaAs HBT. (**a**) The doping and thickness of layers. (**b**) The process technology (The red, green, and blue regions are emitter, base, and collector ohmic contacts, respectively. The yellow area is the metal deposited so that the electrodes are at the same height).

In the experiment, the irradiation was performed using 10 MeV protons with a fluence of 1×10^{12} p/cm^2 at the Institute of Heavy Ion Physics, Peking University. All terminals of the device were floated. The 10 MeV protons were incident vertically from the front of the device and could penetrate it, as simulated by the TRIM tool. After irradiation, the samples were self-annealed for 100 days. The DC and AC characteristics were measured at room temperature (T = 300 K) for pre-irradiation, post-irradiation and post-annealing. In the next section, the properties of InP HBT will be discussed with model parameter analysis, including the forward- and reverse-mode Gummel characteristics, resistances, capacitances, and frequency characteristics.

3. Model and Analysis

Numerous models have been proposed to describe and predict the properties of bipolar transistors, including the EM model, GP model, VBIC model, MEXTRAM model, HICUM model, and Agilent HBT model. Research studies have indicated that the Agilent HBT model could characterise the properties of InP HBT accurately [24–27]. Consequently, this model would be selected in this paper to examine the influence of proton irradiation on InP HBTs.

Figure 2 shows the schematic diagram of the Agilent HBT model. Notably, the diagram does not include the parasitic elements, such as the extrinsic capacitances C_{PBE}, C_{PBC} and C_{PCE}, as well as the inductances L_{PB}, L_{PE} and L_{PC}. These elements are typically eliminated using the Open-Short de-embedding technique. Due to the triple mesa process and the high doping density in the base region of InP HBT, the extrinsic base and collector current and charge are smaller than the intrinsic values. Consequently, the extrinsic base and collector current and charge could be disregarded, and the ABEX and ABCX parameters are set to their default values [20]. I_{BEi}, Q_{BEi}, I_{BCi} and Q_{BCi} represent the intrinsic base and collector current and charge, while I_{CE} is the collector–emitter current. Additionally, R_{Bx}, R_{Bi}, R_{Cx}, and R_{Ci} correspond to the extrinsic and intrinsic resistances of the base and collector regions, respectively.

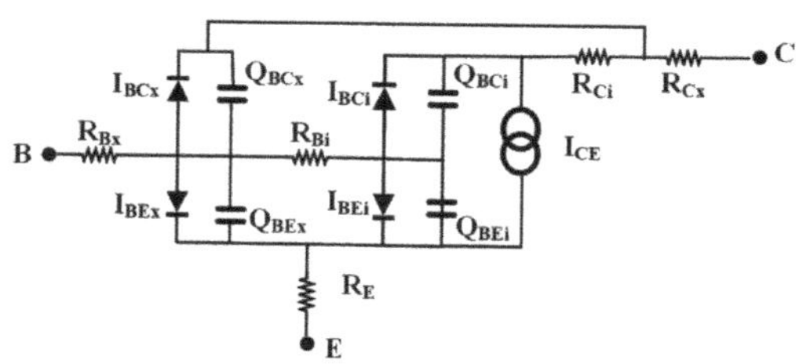

Figure 2. The Agilent HBT model without the parasitic elements.

Parameter extraction begins with the elimination of parasitic elements using the Open-Short structure. Subsequently, the DC parameters are extracted from the forward- and reverse-mode Gummel characteristics, along with the output characteristics. Capacitance, resistance, and frequency characteristics are obtained from the de-embedded S-parameter. Detailed explanations of the model parameters can be found in Reference [28].

To investigate the proton-induced degradation mechanism of InP HBTs, the variation rate (VR) of the model parameters is defined as the ratio of the changes in the parameters after irradiation or annealing to the parameters before irradiation. The positive and negative signs of the variation rate represent an increase and decrease in parameter value, respectively.

3.1. Forward-Mode Gummel Characteristics

The forward-mode Gummel characteristic, an essential input electrical property of bipolar transistors, effectively reflects the properties of the base–emitter (BE) junction. Figure 3a,b show the measured and modelled base and collector currents for the pre-irradiation, post-irradiation and post-annealing stages, with base and collector voltage varying from 0.4 V to 0.9 V. The base and collector current curves are discrete in the low-voltage range due to the electromagnetic interference, vibration, and temperature disturbance. After irradiation, the collector currents remain relatively stable, while the base currents show a significant increase, particularly noticeable when V_{BE} is below 0.75 V.

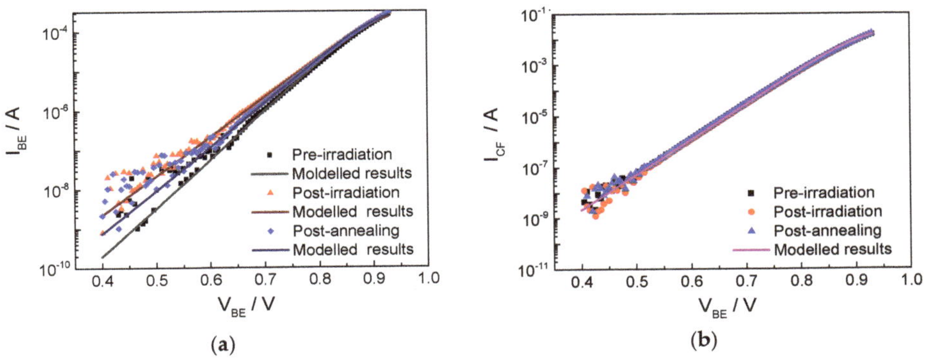

Figure 3. Measured and modelled forward-mode Gummel characteristics for pre-irradiation, post-irradiation and post-annealing of 10 MeV (proton). (**a**) Base current. (**b**) Collector current.

In the Agilent HBT model, the collector current I_{CF} in the forward-mode Gummel characteristic is expressed in Equation (1).

$$I_{CF} = I_S \times \left(\exp\left(\frac{qV_{BEi}}{N_F \times k \times T} \right) - 1 \right) \quad (1)$$

where the parameters *Is* and N_F are the collector saturation current and ideality factor, respectively, which can be obtained from the intercept and slope of the current curves. In addition, k is the Boltzmann constant, *T* is the temperature, V_{BEi} is the voltage applied to the BE junction, and q is the electron charge.

The base–emitter current I_{BE} is given by Equation (2).

$$I_{BE} = {q_{3mod}}^{GKDC} \times I_{SH}\left(\exp\left(\frac{qV_{BEi}}{N_H \times k \times T} \right) - 1 \right) + I_{SE}\left(\exp\left(\frac{qV_{BEi}}{N_E \times k \times T} \right) - 1 \right) \quad (2)$$

where the parameters I_{SH} and N_H are the base–emitter saturation current and ideality factor in the ideal region (the high-voltage region), respectively, while the parameters I_{SE} and N_E are the base–emitter saturation current and ideality factor in the non-ideal region (the low-voltage region), respectively. The method of extracting the model parameters is similar to that of the collector current. The q_{3mod} is an empirical parameter used to model a potential rise in the base current caused by the soft-knee effect. Since this phenomenon is not observed in the measured InP HBTs, the GKDC parameter is set to its default value of 0 [22]. The model parameters of forward Gummel characteristics are extracted and presented in Table 1. As shown in Figure 3, the modelled results demonstrate good agreement with the measured results across a wide voltage range, confirming the effectiveness of the extracted parameters.

In the BE junction, when both the base and collector voltages vary from 0.4 to 0.9 V simultaneously, the BE junction is forward biased while the BC junction is always zero biased. At low voltages, the BE junction is weakly forward biased, causing a small number of electrons diffusing from the emitter region to the base region to spend more time in the space charge region (SCR). This results in the base current at low voltages being primarily dominated by the recombination current in the SCR, which could be characterised by the model parameters I_{SE} and N_E. Furthermore, proton irradiation could exacerbate the non-ideal current behaviour shown in Figure 3a. As with the previous analysis in Ref. [8], the interaction of protons with material atoms in InP HBT generates acceptor traps of In vacancies, which act as recombination centres. These traps could reduce the lifetime of minority carriers and increase the recombination rate in devices. As shown in Table 1, the parameter with maximum VR is I_{SE}, which indicates that a large number of irradiation defects are induced in the BE junction and on the InGaAs surface near the base region. Moreover, the value of N_e is generally between 1 and 2, which could characterise the property of recombination current in the SCR. A value of N_e closer to 2 indicates a higher recombination current. When compared to the parameters N_F and N_H, a higher N_E suggests that the irradiation-induced defects lead to a significant recombination current in the SCR of the BE junction. It is most likely associated with the effect of trap-assisted tunnelling in the SCR and the formation of a surface channel near the base region. The band diagram under forward bias, as shown in Figure 4, is used to analyse the conduction mechanism of the BE junction. Irradiation-induced defects in the SCR of the BE junction not only contribute to the diffusion current and the hot carrier emission current but also directly participate in tunnelling process.

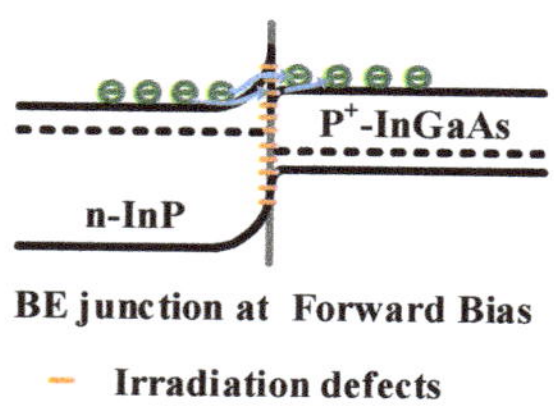

Figure 4. The band diagrams under forward bias (The green symbols represents electrons).

Table 1. The current parameters of forward-mode Gummel characteristics for pre-irradiation, post-irradiation, and post-annealing.

	Pre-Irradiation	Post-Irradiation	VR	Post-Annealing	VR
I_S (A)	6.83×10^{-15}	6.80×10^{-15}	−0.44%	6.81×10^{-15}	−0.31%
N_F	1.22	1.23	0.82%	1.22	0
I_{SH} (A)	2.41×10^{-16}	3.00×10^{-16}	24.48%	2.20×10^{-16}	−8.71%
N_H	1.29	1.35	4.65%	1.30	0.77%
I_{SE} (A)	2.45×10^{-15}	2.50×10^{-13}	10,104.08%	3.30×10^{-14}	1246.94%
N_E	1.41	1.71	21.28%	1.56	10.64%

In the high-voltage region, the BE junction is strongly forward biased, resulting in a narrow space charge region (SCR). Consequently, the recombination current in the SCR of the BE junction can be disregarded. A large number of electrons diffuse from the emitter region to the base region. Some of these electrons recombine with holes in the quasi-neutral bulk base region (NBR), generating the recombination current, while the remaining electrons diffuse to the collector region boundary and are collected by the BC junction. Therefore, the recombination current in the NBR dominates the base current in the high-voltage region, which can be modelled using the parameters I_{SH} and N_H. In Table 1, the middle VR appearing in I_{SH} and N_H suggests that there is a minor amount of displacement damage induced by proton irradiation in the bulk base region compared to that induced in the SCR, resulting in a slight increase in the recombination current in the high-voltage region.

Based on the analysis of the base current, the increase in recombination current in the base region would theoretically cause a decrease in the total number of carriers collected by BC junction. However, this decrease is not clearly discernible in Figure 3b. Nevertheless, when examining the model parameters extracted in Table 1, it is found that the negative value of VR of collector saturation current I_S implies a minor reduction in the collector current. Furthermore, the minimal VRs of collector saturation current I_S and ideality factor

N_F indicate that the collection charge capability of BC junction is nearly saturated, with the collector current being marginally influenced by proton irradiation. Therefore, it is evident that the model parameters offer a more detailed insight into the effects of irradiation on the devices.

After annealing, the VRs of the parameters decrease significantly compared to the post-irradiation stage. This suggests that some defects generated by displacement damage are recombined after a period of annealing, leading to a slight recovery of the degraded characteristics. However, the remaining defects form stable traps that will have a long-term impact on the DC and AC characteristics of the device. Therefore, the degraded characteristics of post-annealing do not fully recover to the pre-irradiation state.

The current gain ($\beta = I_{CF}/I_{BE}$) of the InP HBT is shown in Figure 5. It is evident that the gain decreases after irradiation, which is also attributed to the increase in the recombination current of the BE junction induced by irradiation damage.

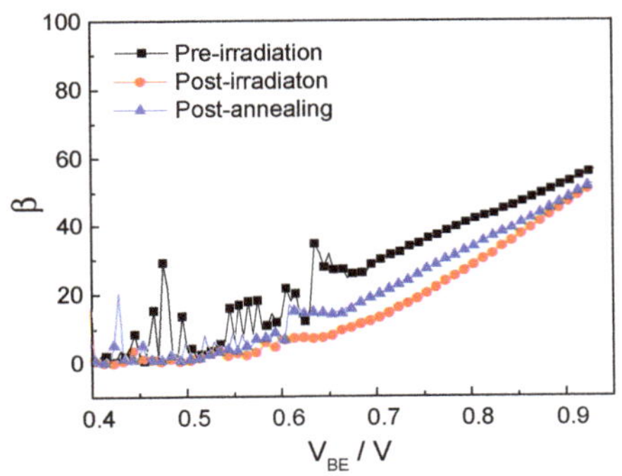

Figure 5. Current gain (β) for pre-irradiation, post-irradiation, and post-annealing of 10 MeV (proton).

3.2. Reserve-Mode Gummel Characteristics

For NPN bipolar transistors, the electrical property of BC junction can be reflected by the reverse-mode Gummel characteristic. The measured and modelled results for pre-irradiation, post-irradiation, and post-annealing with the collector voltage varying from −0.2 V to −0.8 V, and the emitter and base electrodes grounded are shown in Figure 6a,b. The good agreement between the modelled results and experimental data across a wide range in Figure 6 confirms the accuracy of the extracted parameters.

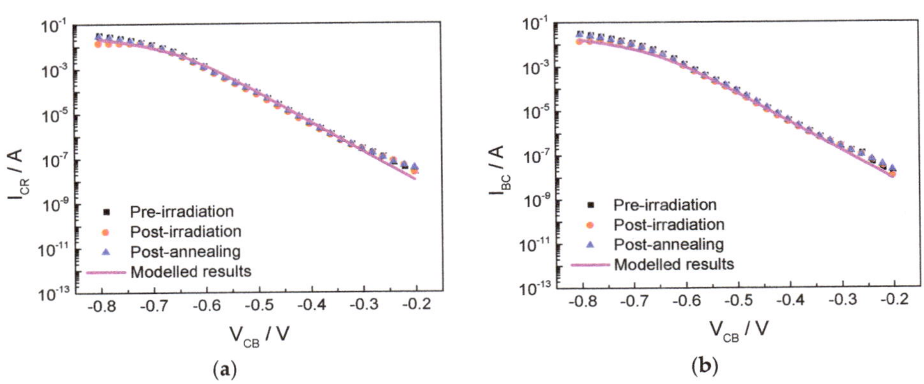

Figure 6. Measured and modelled reverse-mode Gummel characteristic for pre-irradiation, post-irradiation, and post-annealing of 10 MeV (proton). (**a**) Collector current. (**b**) Base current.

In the compact model, the reverse collector current I_{CR} and base current I_{BC} are expressed in Equations (3) and (4), respectively.

$$I_{CR} = I_{SR} \times \left(\exp\left(\frac{qV_{BCi}}{N_R \times k \times T}\right) - 1\right) \tag{3}$$

$$I_{BC} = I_{SRH}\left(\exp\left(\frac{qV_{BCi}}{N_{RH} \times k \times T}\right) - 1\right) + I_{SC}\left(\exp\left(\frac{qV_{BCi}}{N_C \times k \times T}\right) - 1\right) \tag{4}$$

where the parameters include the collector saturation current I_{SR}, ideality factor N_R, base–collector saturation current I_{RSH}, ideality factor N_{RH} in the ideal region, as well as the base–collector saturation current I_{SC} and ideality factor N_C in the non-ideal region. V_{BCi} stands for the applied voltage of BC junction. The saturation current and the ideality factor parameters are extracted from the intercept and slope of the different regions, as shown in Table 2.

Table 2. The current parameters of reverse-mode Gummel characteristics for pre-irradiation, post-irradiation, and post-annealing.

	Pre-Irradiation	Post-Irradiation	VR	Post-Annealing	VR
I_{SR} (A)	2.47×10^{-12}	2.47×10^{-12}	0	2.47×10^{-12}	0
N_R	1.20	1.20	0	1.20	0
I_{SRH} (A)	6.72×10^{-16}	6.72×10^{-16}	0	6.72×10^{-16}	0
N_{RH}	1.03	1.03	0	1.03	0
I_{SC} (A)	2.49×10^{-11}	2.49×10^{-11}	0	2.49×10^{-11}	0
N_C	1.31	1.31	0	1.31	0

In the reverse-mode Gummel characteristics, as the collector voltage varies from −0.2 to −0.8 V, the BC junction is forward biased while the BE junction is zero biased. Similar to the base current I_{BE} in forward-mode Gummel characteristics, the base current I_{BC} in the reverse characteristics is primarily determined by the recombination current in the space charge region of the BC junction (which can be characterised by parameters I_{SC} and N_C) and the recombination current in the quasi-neutral bulk base region (related to I_{RSH} and N_{RH}). As seen in Table 2, the VRs of the model parameters are all zero in the reverse Gummel characteristics, indicating a minimal impact of proton irradiation on the BC junction. This implies that the displacement damage induced in the SCR of the BC junction can be disregarded.

As the voltage of BC junction increases, the collector and base currents gradually saturate, which is attributed to the collector resistance caused by the low doping of the collector region. The voltage drop across the collector resistance in the high-voltage region limits the increase in current, known as the high-level injection effect. Compared to the pre-irradiation and post-annealing state, both the base and collector current decrease slightly after irradiation in the high-voltage region, which may be associated with the collector resistance.

In the model, the extrinsic emitter resistance R_E, base resistance R_{BX} and collector resistance R_{CX} can be determined using the Open-Collector method, which involves extrapolating the curves of the real part of the Z-parameters as expressed in Equations (5)–(7) [29,30]. The intrinsic resistance $R_{CI,}$ R_{BI}, along with the thermal resistance, can be obtained from fitting the output characteristics in Figure 7.

$$R_{bx} = real(Z_{11} - Z_{12})_{1/Ib \to 0} \tag{5}$$

$$R_e = real(Z_{12})_{1/I_b \to 0} \tag{6}$$

$$R_{cx} = real(Z_{22} - Z_{21})_{1/Ib \to 0} \tag{7}$$

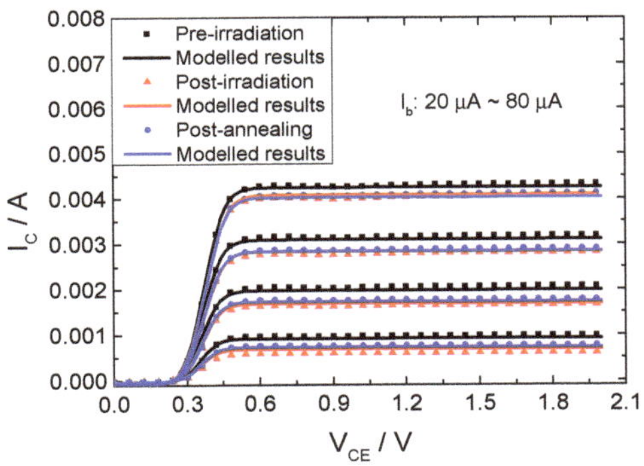

Figure 7. Measured and modelled output characteristics before irradiation, after irradiation, and after annealing of 10 MeV (proton).

As shown in Table 3, the intrinsic and extrinsic collector resistances R_{CI} and R_{CX} show the most significant changes compared to the other resistances with rates of 4.78% and 25.77%, respectively. Due to its large area and low doping, the collector region is susceptible to proton irradiation. As a result, irradiation-induced damage results in a slight increase in the collector resistance in the quasi-neutral bulk collector region. Therefore, due to the effect of the high-level injection, the increased collector resistance is responsible for the slight decrease in base and collector current in the high-voltage region after irradiation, as shown in Figure 6.

Table 3. The resistance parameters of resistances for pre-irradiation, post-irradiation, and post-annealing.

	Pre-Irradiation	Post-Irradiation	VR	Post-Annealing	VR
R_{BX} (Ω)	0.31	0.32	3.22%	0.31	0
R_{BI} (Ω)	0.1	0.1	0	0.1	0
R_{E} (Ω)	2.1	2.1	0	2.1	0
R_{CX} (Ω)	3.88	4.88	25.77%	3.88	0
R_{CI} (Ω)	2.09	2.19	4.78%	2.09	0
R_{th1} (Ω)	0.1	0.1	0	0.1	0
R_{th2} (Ω)	0.1	0.1	0	0.1	0

Figure 7 shows the comparison between measured and modelled output characteristics at base currents of 20 μA, 40 μA, 60 μA, and 80 μA for pre-irradiation, post-irradiation, and post-annealing, respectively. The strong consistency between measured and modelled results, with an error within 5%, validates the accuracy of the DC model and also demonstrates the effectiveness of the analysis. Moreover, at a fixed base current, the increased carrier recombination in the base region due to irradiation damage results in fewer carriers being collected by the BC junction. Consequently, the collector current decreases after irradiation, with the model parameter analysis of the forward Gummel characteristics.

3.3. Capacitance Characteristics

The characteristics of the defects can also be evaluated by observing their influence on the capacitance. When the BE and BC junction are reverse biased or weakly positive biased (with the base voltage varying from −3 V to 0.4 V), the S-parameters are measured. Then, the junction capacitance can be calculated from the imaginary part of the Y-parameters (converted from S-parameters) in Equations (8) and (9) [31,32].

$$C_{be} = \frac{imag(Y_{11} + Y_{12})}{\omega} \quad (8)$$

$$C_{bc} = -\frac{imag(Y_{12})}{\omega} \tag{9}$$

Figure 8a,b show the capacitance of the base–emitter junction (C_{BE}) and base–collector junction (C_{BC}) before irradiation, after irradiation, and after annealing. The great agreement between the measured and modelled results for both the BE and BC junctions confirms the accuracy of the extracted model parameters. As shown in Figure 8, the C_{BE} increases after irradiation and then decreases slightly after annealing, while the C_{BC} is hardly affected by proton irradiation. This differs from the capacitance behaviour of InP HBT under gamma irradiation, as reported in Reference [22], where both C_{BE} and C_{BC} show an increase.

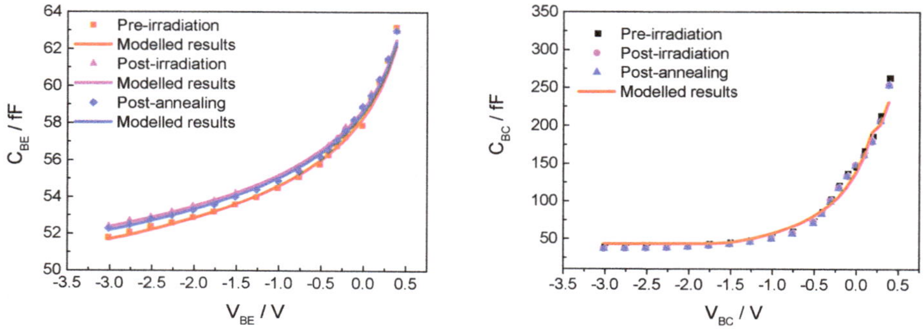

(a) (b)

Figure 8. Measured and modelled base–emitter and base–collector junction capacitance for pre-irradiation, post-irradiation, and post-annealing. (**a**) BE junction capacitance (C_{BE}); (**b**) BC junction capacitance (C_{BC}).

In the compact model, the equations for the junction capacitances can be found in the Reference [28]. C_{BE} is characterised by several parameters, namely the zero-biased capacitance C_{je}, maximum value in forward bias C_{emax}, built-in voltage V_{je}, grading factor M_{je}, punch-through voltage V_{pte}, and grading factor beyond punch-through M_{jer}. Similarly, C_{BC} also has analogous parameters, including C_{jc}, C_{cmax}, V_{jc}, M_{jc}, V_{ptc} and M_{jcr}, which are listed in Table 4.

Table 4. The capacitance parameters for pre-irradiation, post-irradiation, and post-annealing.

	Pre-Irradiation	Post-Irradiation	VR	Post-Annealing	VR
C_{je} (F)	1.672×10^{-14}	1.677×10^{-14}	0.3%	1.674×10^{-14}	0.12%
C_{emax} (F)	6.06×10^{-14}	6.06×10^{-14}	0	6.06×10^{-14}	0
V_{je} (V)	0.64 V	0.64 V	0	0.64 V	0
M_{je}	0.78	0.78	0	0.78	0
V_{pte}	0.11	0.11	0	0.11	0
M_{jer}	0.070	0.067	−4.28%	0.068	−2.94%
C_{jc} (F)	1.35×10^{-13}	1.35×10^{-13}	0	1.35×10^{-13}	0
C_{cmax} (F)	2.01×10^{-13}	2.01×10^{-13}	0	2.01×10^{-13}	0
V_{jc} (V)	0.67 V	0.67 V	0	0.67 V	0
M_{jc}	0.96	0.96	0	0.96	0
V_{ptc}	2.22	2.22	0	2.22	0
M_{jcr}	0.01	0.01	0	0.01	0

Apart from C_{je} and M_{jer}, no changes were observed in the other parameters. This implies that the irradiation-induced degradation mainly takes place in the BE junction rather than the BC junction, which is consistent with the analysis of the parameters in the forward and reverse Gummel characteristics. In the BE junction, the defects induced by proton irradiation could capture or release carriers with the changes in applied voltage, which is equivalent to the effect of charging or discharging, thus resulting in an increase in C_{BE} after irradiation [33]. Furthermore, the absolute rate of VR for M_{jer} is higher compared to C_{je}. M_{jer}, defined as the grading factor beyond punch-through, is the parameter that describes the slope of the punch-through region for the base–emitter junction. It can not only reflect changes of C_{BE}, but is also associated with the tunnel phenomenon, confirming the trap-assisted tunnelling mechanism in the recombination current of the BE junction.

In addition, the VRs of the model parameters decrease after annealing, indicating a recovery of the defects in the BE junction compared to the post-irradiation state. However, this is different from the capacitance property of GaAs HBT under gamma irradiation in Reference [21], where further degradation was reported after annealing compared to the post-irradiation condition.

3.4. Frequency Characteristics

The cut-off frequency (F_t) is the frequency at which the current gain drops to unity in a common-emitter transistor, which is a crucial parameter in characterising the AC performance of high-frequency devices. To determine F_t, the S-parameters and collector current are measured as the base current varies from 10 µA to 150 µA with the collector voltage fixed at 1.7 V. By extrapolating the $H(2,1)$ curves derived from S-parameters at different collector currents, the cut-off frequency can be calculated. The measured and modelled results of the cut-off frequency for pre-irradiation, post-irradiation, and post-annealing are shown in Figure 9a. It is observed that the cut-off frequency exhibits more pronounced variations with increasing collector current.

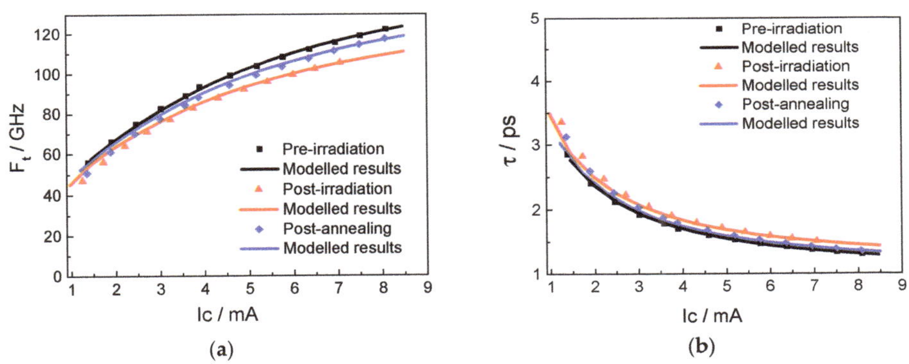

Figure 9. Measured and modelled results for pre-irradiation, post-irradiation, and post-annealing of 10 MeV (proton). (**a**) Cut-off frequency (F_t). (**b**) Transit time (τ).

Actually, the cut-off frequency is primarily determined by the transit time in the devices, as expressed in Equation (10). A longer transit time results in a lower cut-off frequency. The total transit time (τ) is the sum of several individual transit times, including the charging time in the emitter region (τ_e), the transit time in the base region (τ_b), the transit time in the depletion region of the BC junction (τ_{sc}) and the charging time in the collector region (τ_c), as described in Equations (11)–(14). The cut-off frequency can be improved effectively by optimising these transit times.

$$f_T = \frac{1}{2\pi\tau} = \frac{1}{2\pi(\tau_e + \tau_b + \tau_c + \tau_{sc})} \tag{10}$$

$$\tau_e = \frac{kT}{qI_C}(C_{BE} + C_{BC}) \tag{11}$$

$$\tau_b = \frac{T_B{}^2}{2D_{nB}} \tag{12}$$

$$\tau_{sc} = \frac{T_{dep}}{2v_{eff}} \tag{13}$$

$$\tau_c = (R_E + R_C)\cdot C_{BC} \tag{14}$$

where T is the temperature, C_{BE} and C_{BC} are the BE and BC junction capacitance, respectively, T_B is the thickness of the base region, D_{NB} is the electron diffusion coefficient in the base region, T_{dep} is the width of the BC junction depletion region, v_{eff} is the electron average rate in the depletion region of BC junction, and R_E and R_C are the emitter and collector resistances, respectively.

Nevertheless, in the compact model, three types of transit time are considered. The base region (τ_b) and BC junction (τ_{sc}) transit time correspond to Equations (12) and (13), respectively. Additionally, if the cut-off frequency decreases at the high current injection, the excess delay due to the Kirk effect (τ_{ke}) must be considered; otherwise, default parameters would be used. Detailed equations and descriptions of these transit time parameters can be found in Reference [26]. These transit times, along with the transit time caused by capacitance and resistance in Equations (11) and (14), collectively determine the total transit time (τ), as shown in Figure 9b. The error between the measured and modelled results is within 3%, validating the accuracy of the AC model. The extracted model parameters are listed in Table 5.

Table 5. The transit time parameters for pre-irradiation, post-irradiation, and post-annealing.

	Pre-Irradiation	Post-Irradiation	VR	Post-Annealing	VR
T_{fb} (S)	4.60×10^{-13}	5.40×10^{-13}	17.4%	5.10×10^{-13}	10.9%
T_{fc0} (S)	1.33×10^{-11}	1.33×10^{-11}	0	1.33×10^{-11}	0
T_{cmin} (S)	3.20×10^{-14}	3.20×10^{-14}	0	3.20×10^{-14}	0
I_{tc} (A)	7.60×10^{-4}	7.60×10^{-4}	0	7.60×10^{-4}	0
I_{tc2} (A)	7×10^{-4}	7×10^{-4}	0	7×10^{-4}	0
V_{tc0lnv}	3	3	0	3	0
V_{tr0} (V)	2	2	0	2	0
V_{mx0} (V)	0.5	0.5	0	0.5	0
V_{tmin} (V)	1.2	1.2	0	1.2	0

As seen in Table 5, it is evident that only the base transit time parameter T_{fb} exhibits significant changes before and after irradiation. For a heavily doped base, it is reasonable to approximate that the transit time in the base region (τ_b) is a constant at low and medium currents (below Kirk effect), which could be modelled by the parameter T_{fb} [26]. As analysed in the forward-mode Gummel characteristics, the changes in parameters I_{SH} and N_H before and after irradiation suggest that the amount of displacement damage is induced in the quasi-neutral bulk base region (NBR). This could lead to an increased carrier recombination rate and a decreased diffusion coefficient (D_{NB}) in the NBR, resulting in an increase in the base transit time (τ_b) according to Equation (12). In addition, based on the analysis of reverse-mode characteristics, the impact of proton irradiation on the collector junction can be neglected. Therefore, the VR of the related model parameters (T_{fc0}, T_{cmin}, I_{tc}, I_{tc2}, V_{tc0lnv}, V_{tr0}, V_{mx0} and V_{tmin}) are zero in Table 5.

Finally, the proton irradiation, the increase in collector resistance, capacitance of the BE junction, as well as the base transit time (τ_b), collectively result in a reduction in the cut-off frequency.

4. Conclusions

The DC and AC characteristics in InP HBT have been studied and modelled using a compact model for pre-irradiation, post-irradiation, and post-annealing. The degradation mechanism could be reflected by the change in model parameters. In the forward and reverse Gummel characteristics, the parameter with the maximum variation rate is I_{SE}, which is related to the recombination current around the BE junction. It indicates that the space charge region of the BE junction and the InGaAs surface near the base region are more significantly affected by proton irradiation. The change in the I_{SH} and N_H parameters before and after irradiation suggests that a minor amount of displacement damage is induced in the quasi-neutral bulk base region (NBR). Based on the previous analysis, acceptor traps of In vacancies are generated by the collision of protons with material atoms, which act as recombination centres and decrease the minority carriers lifetime, resulting in a significant recombination current in the BE junction.

In the AC characteristics, the radiation degradation around the BE junction results in an increase in the C_{BE} and a decrease in the cut-off frequency after irradiation. The variation rate of parameter M_{jer} validates the trap-assisted tunnelling mechanism in the recombination current of the BE junction. The significant changes in the base region transit time parameter T_{fb} before and after irradiation are consistent with the analysis of degradation in the quasi-neutral bulk base region (NBR).

After annealing, the variation rates of the parameters decrease significantly, which indicates a reduction in defect density compared to post-irradiation. Some unstable defects in the BE junction are recombined and disappear after a period of annealing, while the remaining stable traps will have a long-term influence on the device characteristics.

Finally, the good agreement between the modelled and measured results of DC and AC characteristics confirms the accuracy and effectiveness of the extracted model parameters. This lays the foundation for the establishment of an irradiation compact model to predict device characteristics and also provides a powerful tool to design anti-radiation circuits in future work.

Author Contributions: Conceptualisation, X.Z.; methodology, X.Z.; software, X.Z.; validation, X.Z.; formal analysis, Y.C.; investigation, S.C.; resources, Y.Z.; data curation, X.W.; visualisation, Y.B. and S.S.; supervision, H.W. and F.P.; project administration, X.Z.; funding acquisition, X.Z.; writing—original draft, X.Z.; writing—review and editing, G.T. and Y.Y. All authors have read and agreed to the published version of the manuscript.

Funding: This research was funded by National Natural Science Foundation of China (Grant No. 62104260).

Data Availability Statement: All the data are shown in the paper.

Acknowledgments: The authors would like to acknowledge the Institute of Microelectronics of the Chinese Academy of Sciences for providing test samples and Institute of Heavy Ion Physics of Peking University for providing experimental equipment.

Conflicts of Interest: Author Shaowei Sun was employed by the company LONGi Green Energy Technology Co., Ltd. The remaining authors declare that the research was conducted in the absence of any commercial or financial relationships that could be construed as a potential conflict of interest.

References

1. Urteaga, M.; Griffith, Z.; Seo, M.; Hacker, J.; Rodwell, M.J. InP HBT technologies for THz integrated circuits. *Proc. IEEE* **2017**, *105*, 1051–1067. [CrossRef]
2. Yun, J.; Oh, S.J.; Song, K.; Yoon, D.; Son, H.Y.; Choi, Y.; Huh, Y.M.; Rieh, J.S. Terahertz reflection-mode biological imaging based on InP HBT source and detector. *IEEE Trans. Terahertz Sci. Technol.* **2017**, *7*, 274–283. [CrossRef]
3. Yun, J.; Kim, J.; Rieh, J.S. A 280-GHz 10-dBm Signal Source Based on InP HBT Technology. *IEEE Microw. Wirel. Compon. Lett.* **2017**, *27*, 159–161. [CrossRef]
4. Hansen, D.L.; Chu, P.; Meyer, S.F. Effects of data rate and transistor size on single event upset cross-sections for InP-based circuits. *IEEE Trans. Nucl. Sci.* **2005**, *52*, 3166–3171. [CrossRef]
5. Hansen, D.L.; Marshall, P.W.; Lopez-Aguado, R.; Jobe, R.K.; Carts, M.A.; Marshall, C.J.; Meyer, S.F. A study of the SEU performance of InP and SiGe shift registers. *IEEE Trans. Nucl. Sci.* **2005**, *52*, 1140–1147. [CrossRef]
6. Chu, P.; Hansen, D.L.; Doyle, B.L.; Jobe, K.; Lopez-Aguado, R.; Shoga, M.; Walsh, D.S. Ion-microbeam probe of high-speed shift registers for SEU Analysis-part II: InP. *IEEE Trans. Nucl. Sci.* **2006**, *53*, 1583–1592. [CrossRef]

7. Li, C.H.; Lu, H.L.; Zhang, Y.M.; Liu, M.; Zhao, X.H. Proton-Induced Degradation of InP/InGaAs HBTs Predicted by Nonionizing Energy Loss Model. *IEEE Trans. Nucl. Sci.* **2015**, *62*, 1336–1340. [CrossRef]
8. Zhao, X.H.; Lu, H.L.; Zhao, M.L.; Zhang, Y.M.; Zhang, Y.M. The Study of Deep Level Traps and Their Influence on Current Characteristics of InP/InGaAs Heterostructures. *Nanomaterials* **2019**, *9*, 1141. [CrossRef]
9. Zhao, X.H.; Lu, H.L.; Zhang, Y.M.; Zhang, Y.M.; Wei, Z.C. The impact of laser energy, bias and irradiation positions on single event transients of InP HBT. *J. Phys. D Appl. Physic.* **2020**, *53*, 145104. [CrossRef]
10. Wei, J.N.; He, C.H.; Li, P.; Li, Y.H.; Guo, H.X. Impact of displacement damage on single event transient charge collection in SiGe HBTs. *Nucl. Instrum. Methods Phys. Res. Sect. A Accel. Spectrometers Detect. Assoc. Equip.* **2019**, *938*, 29–35. [CrossRef]
11. Meng, X.T.; Yang, H.W.; Kang, A.G.; Wang, J.L.; Jia, H.Y.; Chen, P.Y.; Tsien, P.H. Effects of neutron irradiation on SiGe HBT and Si BJT devices. *J. Mater. Sci. Mater. Electron.* **2003**, *14*, 199–203. [CrossRef]
12. Petrosyants, K.; Vologdin, E.; Smirnov, D.; Torgovnikov, R.; Kozhukhov, M. Si BJT and SiGe HBT performance modeling after neutron radiation exposure. In Proceedings of the 2011 9th East-West Design & Test Symposium (EWDTS), Sevastopol, Ukraine, 9–12 September 2011; pp. 267–270.
13. Zhang, X.; Li, Y.; Guo, Q.; Feng, J. Low energy proton irradiation effects on InP/InGaAs DHBTs and InP-base frequency dividers. *Young Sci. Forum* **2017**, *10710*, 65–70.
14. Zhang, X.Y.; Zhang, X.; Li, Y.; Guo, Q.; Liao, B.; Yang, S.; Yang, Z.; He, C. Displacement damage effects and anneal characteristic on InP/InGaAs DHBTs. In Proceedings of the 2018 International Conference on Radiation Effects of Electronic Devices (ICREED), Beijing, China, 16–18 May 2018; pp. 1–5.
15. Lawal, O.M.; Liu, S.; Huang, T. Extraction and analysis of gamma irradiated Si BJT SPICE model. *ECS J. Solid State Sci. Technol.* **2019**, *8*, 51. [CrossRef]
16. Petrosyants, K.O.; Kozhukhov, M.V.; Dvomikov, O.V.; Savchenko, E.M.; Budyakov, A.S. SPICE-model of SiGe HBT taking into account radiation effects. In Proceedings of the 2018 Moscow Workshop on Electronic and Networking Technologies (MWENT), Moscow, Russia, 14–16 March 2018; pp. 1–4.
17. Petrosyants, K.O.; Kozhukhov, M.V. TCAD-SPICE Two Level Simulation of Si BJTs and SiGe HBTs Taking into Account Radiation Effects. *Dev. Issues Adv. Microelectron. Nanoelectron. Syst. (MES)* **2017**, *2*, 2–10.
18. Petrosyants, K.; Kozhukhov, M. SPICE model parameters extraction taking into account the ionizing radiation effects. In *Proceedings of the 2014 Proceedings of IEEE East-West Design & Test Symposium, Kiev, Ukraine, 26–29 September 2014*; pp. 1–4.
19. Van Uffelen, M.; Geboers, S.; Leroux, P.; Berghmans, F. SPICE modelling of a discrete COTS SiGe HBT for digital applications up to MGy dose levels. In Proceedings of the 2005 8th European Conference on Radiation and Its Effects on Components and Systems, Cap d'Agde, France, 19–23 September 2005; pp. PF4-1–PF4-5.
20. Zhang, J.; Zhang, Y.; Lu, H.; Zhang, Y.; Yang, S. The model parameter extraction and simulation for the effects of gamma irradiation on the DC characteristics of InGaP/GaAs single heterojunction bipolar transistors. *Microelectron. Reliab.* **2012**, *52*, 2941–2947. [CrossRef]
21. Zhang, J.; Xu, K.; Wang, J.; Liu, M.; Zhang, L.; Liu, B. Modeling and analysis for the effects of gamma irradiation on the DC and AC performance in InGaP/GaAs SHBTs. *Radiat. Eff. Defects Solids* **2020**, *175*, 492–503. [CrossRef]
22. Zhang, J.; Cao, L.; Liu, M.; Liu, B.; Cheng, L. Gamma-induced degradation effect of InP HBTs studied by Keysight model. *Nucl. Sci. Eng.* **2021**, *195*, 173–184. [CrossRef]
23. Jin, Z.; Su, Y.B.; Cheng, W.; Liu, X.Y.; Xu, A.H.; Qi, M. High current multi-finger InGaAs/InP double heterojunction bipolar transistor with the maximum oscillation frequency 253 GHz. *Chin. Phys. Lett.* **2008**, *25*, 3075–3078.
24. Zhang, J.C.; Liu, M.; Wang, J.; Zhang, L.; Liu, B. Modeling of InP HBTs with an Improved Keysight HBT Model. *Microw. J.* **2019**, *62*, 56.
25. Zhang, A.; Gao, J. An improved nonlinear model for millimeter-wave inp hbt including dc/ac dispersion effects. *IEEE Microw. Wirel. Compon. Lett.* **2021**, *99*, 465–468. [CrossRef]
26. Iwamoto, M.; Root, D.E.; Scott, J.B.; Cognata, A.; Asbeck, P.M.; Hughes, B.; D'Avanzo, D.C. Large-signal HBT model with improved collector transit time formulation for GaAs and InP technologies. In Proceedings of the IEEE MTT-S International Microwave Symposium Digest, Philadelphia, PA, USA, 8–13 June 2003; Volume 2, pp. 635–638.
27. Johansen, T.K.; Midili, V.; Squartecchia, M.; Zhurbenko, V.; Nodjiadjim, V.; Dupuy, J.Y.; Riet, M.; Konczykowska, A. Large-Signal Modeling of Multi-Finger InP DHBT Devices at Millimeter-Wave Frequencies. In Proceedings of the International Workshop on Integrated Nonlinear Microwave and mmWave Circuits, Graz, Austria, 20–21 April 2017; pp. 1–3.
28. UCSD HBT Model. Available online: http://hbt.ucsd.edu (accessed on 1 December 2023).
29. Wei, C.J.; Hwang, J.C.M. Direct extraction of equivalent circuit parameters for heterojunction bipolar transistors. *IEEE Trans. Microw. Theory Tech.* **1995**, *43*, 2035–2040.
30. Bousnina, S.; Mandeville, P.; Kouki, A.B.; Surridge, R.; Ghannouchi, F.M. Direct parameter-extraction method for HBT small-signal model. *IEEE Trans. Microw. Theory Tech.* **2002**, *50*, 529–536. [CrossRef]
31. Lee, K.; Choi, K.; Kook, S.H.; Cho, D.H.; Park, K.W.; Kim, B. Direct parameter extraction of SiGe HBTs for the VBIC bipolar compact model. *IEEE Trans. Electron Devices* **2005**, *52*, 375–384. [CrossRef]
32. Samelis, A.; Pavlidis, D. DC to high-frequency HBT-model parameter evaluation using impedance block conditioned optimization. *IEEE Trans. Microw. Theory Tech.* **1997**, *45*, 886–897. [CrossRef]

33. Zhao, X.H.; Lu, H.L.; Zhang, Y.M.; Zhang, Y.M. Impact of proton irradiation with different fluences on the characteristics of InP/InGaAs heterostructure. *Radiat. Eff. Defects Solids* **2019**, *174*, 697–707. [CrossRef]

Disclaimer/Publisher's Note: The statements, opinions and data contained in all publications are solely those of the individual author(s) and contributor(s) and not of MDPI and/or the editor(s). MDPI and/or the editor(s) disclaim responsibility for any injury to people or property resulting from any ideas, methods, instructions or products referred to in the content.

electronics

MDPI

Article

Experimental Study of the Impact of Temperature on Atmospheric Neutron-Induced Single Event Upsets in 28 nm Embedded SRAM of SiP

Shunshun Zheng [1,2], Zhangang Zhang [2,*], Jiefeng Ye [1,2], Xiaojie Lu [2], Zhifeng Lei [2], Zhili Liu [3], Gaoying Geng [3], Qi Zhang [3], Hong Zhang [2] and Hui Li [1]

1 Institutes of Physical Science and Information Technology, Anhui University, Hefei 230601, China; q21201103@stu.ahu.edu.cn (S.Z.); q22201088@stu.ahu.edu.cn (J.Y.); huili@ahu.edu.cn (H.L.)
2 Science and Technology on Reliability Physics and Application of Electronic Component Laboratory, China Electronic Product Reliability and Environmental Testing Research Institute, Guangzhou 510610, China; 202121551427@smail.xtu.edu.cn (X.L.); leizhifeng@ceprei.com (Z.L.); 201931000121@smail.xtu.edu.cn (H.Z.)
3 The 58th Institute of China Electronics Technology Group Corporation, Wuxi 214035, China; liuzhl@cetccks.com (Z.L.); genggy@cetccks.com (G.G.); zhangqi@cetccks.com (Q.Z.)
* Correspondence: zhangangzhang@163.com

Abstract: In this paper, the temperature dependence of single event upset (SEU) cross-section in 28 nm embedded Static Random Access Memory (SRAM) of System in Package (SiP) was investigated. An atmospheric neutron beam with an energy range of MeV~GeV was utilized. The SEU cross-section increased by 39.8% when the temperature increased from 296 K to 382 K. Further Technology Computer Aided Design (TCAD) simulation results show that the temperature has a weak impact on the peak pulse current, which is mainly caused by the change of bipolar amplification effect with temperature. As the temperature increases, the critical charge of the device decreases by about 4.8%. The impact of temperature on the SEU cross-section is determined competitively by the peak pulse current and the critical charge. The impact of temperature on critical charge is expected to become more severe as the feature size is further advanced.

Keywords: neutron; SEU; temperature; SRAM; bipolar amplification; critical charge

1. Introduction

With the rapid development of microelectronics technology, the feature size and operating voltage of electronic devices are decreasing, the operating frequency is increasing, and the number of devices per unit area is also increasing, which results in the single event effect (SEE) induced by atmospheric neutrons in aviation and ground electronic systems becoming more and more significant, and the reliability problems caused by this have attracted a lot of attention [1]. In recent years, airplane malfunctions and supercomputing data failures induced by atmospheric neutrons have become commonplace. In the aviation environment, atmospheric neutron-induced SEE can lead to system state errors and data upset, thus posing a considerable threat to the reliability of electronic devices, which can, in turn, jeopardize human life safety [2–5].

In the aviation environment, electric devices are exposed to atmospheric neutron radiation and temperature extremes (temperatures on the sunlit side of the device can be over 400 K, while temperatures on the shaded side can be as low as 90 K). Temperature affects the carrier transport process and the collisional ionization process inside the semiconductor device, which in turn leads to changes in the charge collection process inside the device [6]. Simulation work on Si p+/n/n+ diodes shows a significant temperature dependence of the transient current induced by heavy ions [7,8]. Liu and his coworkers conducted SEU irradiation of C^{12} heavy ions for 0.5-micron partially depleted Silicon-On-Insulator (SOI)

SRAM devices. They found that the single event cross-section of the SRAM devices increased by 98% when the operating temperature was increased from 293 K to 388 K [9]. Cai and his coworkers investigated the temperature dependence of the SEU cross-section of commercial 0.15-micron Thin Film Transistor (TFT) SRAMs in the temperature range of 215–353 K. It was found that the SEU cross-section hardly varied with the temperature in the saturated region of the cross-section. In contrast, in the rising region of the cross-section, the SEU cross-section increased by 257% at the incidence of the Cl particles [10]. TCAD simulation results of charge sharing in 130 nm Complementary Metal Oxide Semiconductor (CMOS) technology devices show an increase in charge sharing collection with increasing temperature [11]. M. Bagatin and his coworkers irradiated commercial SRAMs with thermal and wide-spectrum neutron beams, studying the impact of temperature on the soft error rate. The experimental data, simulations, and analytical modeling show that, depending on the device, temperature may or may not increase the soft error sensitivity by a modest amount [12]. J. Cao et al. investigated the variation of SET pulse widths (irradiation by alpha particles) on 7 nm bulk FinFET D-FF in the temperature range of 300–360 K. The experimental results show that the SET pulse width at the 7 nm node is significantly affected by temperature [13]. Cai et al. investigated the variation of SET pulse width of 16 nm bulk FinFET inverter chain at different temperatures. The results show that an increase in temperature enhances the parasitic bipolar effect of the bulk FinFET technology, leading to an increase in the SET pulse width [14]. However, due to the lack of available spallation neutron sources, the above-mentioned research results on the impact of temperature on the SEE are based on heavy ions, and there is a lack of research on the case of the impact of temperature on the SEE of SRAM devices under atmospheric neutron radiation conditions. Unlike heavy ions, neutrons are not electrically charged, but when incident on the device, a nuclear reaction produces secondary particles, which react with the device to cause SEE [15,16]. The SEE induced by neutrons consists of the following two aspects: thermal neutrons react with B^{10} to produce alpha particles and heavy ions [17]; high-energy and medium-energy neutrons react with materials such as Si in the device to produce recoil heavy nuclei, protons, etc. [18].

In this paper, an Atmospheric Neutron Irradiation Spectrometer (ANIS) [19], which was jointly constructed by the China Electronic Product Reliability and Environmental Testing Research Institute (CEPREI) and China Spallation Neutron Source (CSNS), was used to conduct variable experiments on 28 nm static random memories under different temperature conditions and investigate the variation of SEU in 28 nm embedded SRAM of SiP in the temperature range of 296–382 K. The transient pulse currents are also simulated using TCAD simulation software to analyze their physical mechanisms.

2. Experiments at the ANIS

2.1. Experimental Device

Memory is an important part of microelectronic devices. Atmospheric neutron SEE was investigated using SRAM embedded in SiP as test board. The SRAM adopts the CMOS process, and its core memory cell is a 6-transistor structure with a supply voltage of 0.9 V and a feature size of 28 nm. The testing capacity of SRAM is 1020 KB.

2.2. Experimental Platform

JEDEC standards refer to a series of standards established by the Joint Electron Device Engineering Council (JEDEC) to regulate semiconductor products, technologies, and testing methods [20]. According to the JEDEC Standard JESD89A, the integrated neutron flux at the terrestrial level is equal to 20 $n/(cm^2 \cdot h)$ in the energy range > 1 MeV. This value is referenced to New York City at sea level, under conditions of average solar activity. Figure 1 shows ANIS's simulated differential neutron spectrum at 20 m, with 100 kW of proton beam power. The Joint Electron Device Engineering Council (JEDEC) neutron spectrum is also presented for comparison. The differential spectrum matches the JEDEC standard well in the full energy range. In addition, the neutron flux is approximately 10^9 greater than JEDEC standards. Therefore, it is very suitable for the study of atmospheric neutron SEE.

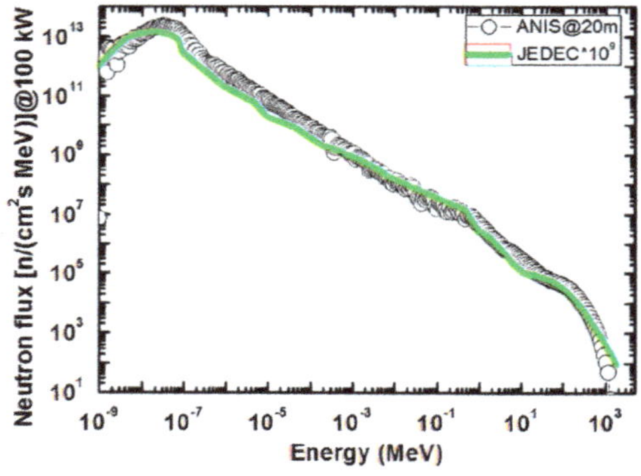

Figure 1. The differential neutron spectrum compared with the JEDEC standard [19].

The experimental site consists of two parts: the irradiation room, and the control room, where the tester writes, reads, and detects the experimental data of the test system through the host computer. The irradiation and control rooms are connected via Ethernet, as shown in Figure 2.

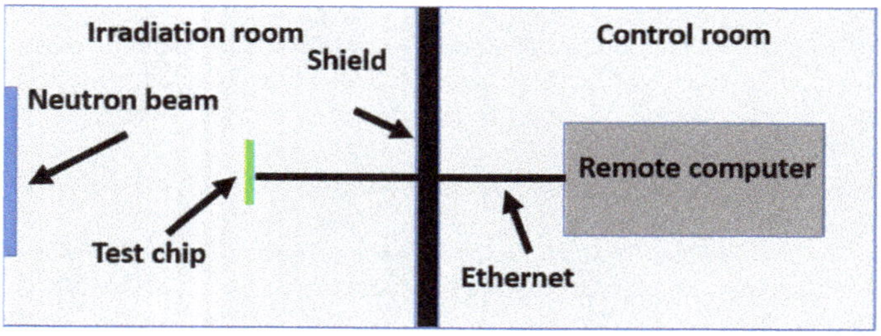

Figure 2. Laboratory layout diagram.

The experiments were conducted at room temperature as well as high temperature. The test board was placed in a thermal cycling chamber to heat it during the experiment. Considering the memory's working temperature range, the temperature points selected for the thermal cycling chamber were 280, 300, 330, and 360 K, respectively. The control panel of the thermal cycling chamber is outside the irradiation room, so the temperature of the thermal cycling chamber can be controlled directly from the outside. Considering the heat of the chip, in order to detect the temperature of the test board in real-time, one side of the PT100 temperature sensor was pasted on the surface of the chip, and the other side was connected to the PID digital display temperature controller. The changes in the temperature figures on the digital display temperature controller can be observed from the monitoring of the control room. During the experiment, the chip's temperature was 296, 322, 351, and 382 K. When the temperature of the chip is kept constant, the experiment can be started. Figures 3 and 4 show the experimental field diagram and the thermal chamber.

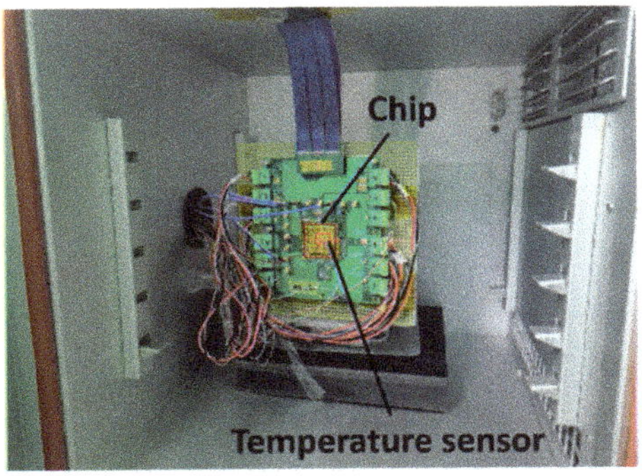

Figure 3. Experimental field diagram.

Figure 4. Thermal chamber.

2.3. *Experimental Method*

The size of the neutron beam used in this experiment is 6 cm $\times$ 10 cm, and the flux rate of neutrons is about 0.85×10^6 n/(cm^2·s). During the experiment, the test board always faced the neutron source, and the neutrons in the experiments were vertically incident. Data 0xA5 was written during the experiment, and to ensure that the experimental data had reliability and ideal statistics, measurements were taken 2–3 times under the same conditions. The experiment was stopped when the number of errors reached 100 or more. The SEU detection includes static and dynamic tests; in order to detect whether the upsets occur in real-time, this experiment chose dynamic detection. The experimental data were written to the memory unit before irradiation, and the data in the memory were read back at intervals (Reading data every second). The number of error bits was counted by comparing the read-out data with the original data written one by one. The chip's current was monitored in real-time during the experiment to avoid single event latch-up and burning the chip. The electric threshold was set. When the current of the chip exceeded the electric threshold, the single event latch-up was considered to have occurred. Then, it was necessary to repower up the daughter board and rewrite the experimental data.

2.4. *Experimental Result*

The experimental data are shown in Table 1.

Table 1. Experimental data.

Temperature	Time	Flux	SEU Number	SEU Cross-Section
296 K	1680 s	1.428×10^9 n/cm^2	125	1.05 cm^2/bit
322 K	2005 s	1.704×10^9 n/cm^2	167	1.17 cm^2/bit
351 K	2242 s	1.906×10^9 n/cm^2	213	1.338 cm^2/bit
382 K	1803 s	1.533×10^9 n/cm^2	188	1.468 cm^2/bit

The SEU cross-section of SRAM is calculated as follows:

$$\sigma_{SEU} = \frac{N_{SEU}}{C\varphi} \tag{1}$$

where σ_{SEU} is the SEU cross-section, N_{SEU} is the number of upsets measured at the irradiated chip, C is the total capacity of the measured SRAM to store data, and φ is the total fluence of the atmospheric neutron. According to data provided by ANIS, the flux rate of atmospheric neutrons is 0.85×10^6 n/(cm^2·s). Figure 5 shows the variation of the SRAM SEU cross-section under different temperature conditions. When the temperature increases from 296 K to 382 K, the cross-section of SRAM increases from 1.05×10^{-14} cm^2/bit to 1.468×10^{-14} cm^2/bit, an increase of 39.8%. From the experimental results, the relationship between the upset cross-section and temperature appears to be approximately linear. Therefore, we conducted a linear fit and obtained the angular coefficient σ for the linear fit, which is 4.86×10^{-17} cm^2/bit/K in this paper. The linear energy transfer (LET) value of secondary particles induced by atmospheric neutrons is very small, mainly causing soft errors, and no latch-up events were observed during the experiment, so it is considered that the experimental devices are not sensitive to latch-up under neutron irradiation. The SEU cross-section of the SRAM versus temperature depends on the transient pulse current and the electrical characteristics of the device, and the relationship between temperature and cross-section variation is approximately linear, which will be explained in the next section.

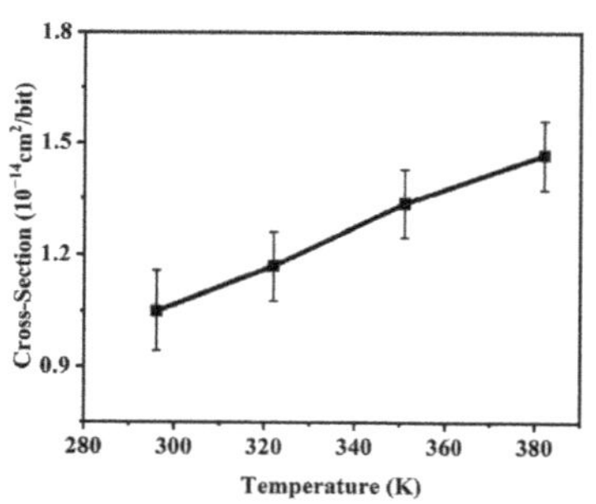

Figure 5. Temperature dependence of SEU cross-section.

3. Discussion

In general, the impact of temperature on SEU depends firstly on the amount of charge deposited within the effective collection depth and secondly on the charge collection process as well as the electrical characteristics of the device, which are all functions of temperature [13,21]. The temperature changes the forbidden bandwidth of the material, which affects the amount of charge deposited by irradiated particle incidence in the adequate collection depth range. The physical mechanisms of charge collection include drift, diffusion, and bipolar amplification. Temperature changes the amount of charge collected by the three mechanisms [22]. Changing temperature changes the upset voltage of the device, and a decrease in the upset voltage reduces the critical charge of the device, thus affecting its SEU cross-section [10].

3.1. Temperature Dependence of Deposited Charge

An increase in temperature leads to a narrowing of the band-gap of Si, and a decrease in the band gap means that more electron-hole pairs are produced for the same linear energy transfer (LET) value of heavy ions. Therefore, the impact of temperature on the electron-hole pairs deposited by heavy ions into the device is mainly reflected in its impact on the band-gap. The variation of band-gap with temperature can be described by the following equation [23]:

$$E_g(T) = E_g(0) - \frac{\alpha T^2}{T + \beta} \tag{2}$$

where T is the absolute temperature, $E_g(0)$ is the band-gap energy at 0 K, and α and β are material parameters. For silicon, α is 4.3×10^{-4} eV/K, and β is 636 K by default. We used default α and β values for all simulations, and the band-gap decreases from 1.1251 to 1.1018 eV when the temperature rises from 296 K to 382 K. According to [24], the average energy expended per ionized electron-hole pair in silicon is

$$\varepsilon = 2.2E_g(T) + 0.96E_g^{3/2}(T)exp(0.75E_g(T)/T) \tag{3}$$

As temperature increases from 296 K to 382 K, the average energy expended per ionized electron-hole pair in silicon decreases from 3.62 eV to 3.54 eV, suggesting that depositing the same energy increases the Coulomb charge by 2.26%.

3.2. Temperature Dependence of Charge Collection Efficiency

Changes in collected charge leads to variations in pulse current, which consists of two parts: pulse current peak and pulse current duration. Changes in the pulse current peak and duration will change the device's LET threshold, thereby affecting the upset cross-section of the device. The total charge collected is the sum of the drift, diffusion, and bipolar amplification effect, and the impact of temperature on charge collection is reflected in the impact on each of these mechanisms [23]. Increased temperature leads to decreased mobility, and as a result, less charge is collected by drift. The drift charge determines the value of the peak pulse current. A change in temperature leads to a change in the diffusion length and change the charge collected by diffusion. The diffusion charge mainly determines the transient pulse current duration. An increase in temperature causes the bipolar amplification effect to be more significant, causing more charge to be collected. The bipolar amplification charge can affect the peak pulse value and the pulse current duration. The impact of temperature variation on the pulse current characteristics is investigated below from the above three points.

In this paper, Synopsys' Sentaurus TCAD tool was used to analyze the single event charge collection mechanism of 28 nm MOSFET. The simulation model of the 3D device is given in Figure 6. The source/drain doping concentration of the device is 1×10^{19} cm^{-3}, and the well doping concentration is 1×10^{17} cm^{-3}. The relevant dimensional parameter values of the device are shown in Table 2. The single event transient response is most sensitive when the transistor is in the off state, so all simulations were performed with the device in the off state [25]. The model was calibrated to the SPICE model before the simulation started to make the established model closer to the physical characteristics of the actual device.

The following physical models were used: (1) Fermi-Dirac statistics, (2) band-gap narrowing effect, (3) doping-dependent Shockley-Read-Hall (SRH) recombination and Auger recombination, (4) temperature, doping, electricfield and carrier-carrier-scattering impact on mobility, and (5) incident heavy ions were modeled using a Gaussian radial profile with a characteristic 1/e radius of 10 nm. A hydrodynamic model was used for carrier transport. Unless otherwise specified, default models and parameters provided by Sentaurus TCAD vN-2017.09-SP1 (Synopsys, Mountain View, CA, USA) were used.

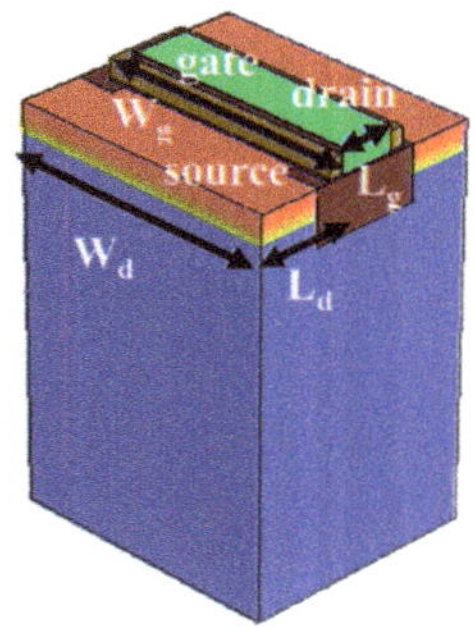

Figure 6. 3D physical model of the device.

Table 2. Structural Parameters of Device.

Dimension Parameter/nm	Value
Drain width: W_d	140
Drain length: L_d	38
Gate width: W_g	28
Gate length: L_g	120

Heavy ions were selected to incident vertically at the center of the drain region of the N-type device, and the temperature of the device was chosen to be 296, 322, 351, and 382 K, respectively. The LET values of secondary ions produced by the interaction of atmospheric neutrons with materials are relatively small. If the LET value is too small, it will not cause upset in the device. The number of ions with larger LET values is very limited, resulting in minimal impact on the devices. Therefore, ions with LET values of 1, 5, and 10 MeV·cm^2/mg are selected for simulation. In studying the impact of bipolar amplification on the transient current of devices, the source is not doped. Therefore, the drain and channel can be approximated as a diode. Since the source electrode is not doped, it cannot inject electrons into the channel, eliminating the bipolar amplification effect. Therefore, the transient current of the diode represents drift diffusion current [22]. Subtracting the total charge from the drift-diffusion collected charge to get the bipolar amplification collected charge.

3.2.1. Temperature Dependence of Peak Pulse Current

To visualize the relationship between peak pulse current and temperature, this paper plots the peak pulse current and the temperature as a curve in Figure 7. When the temperature increased from 296 K to 382 K, the peak pulse current with LET of 1, 5, and 10 Mev·cm^2/mg decreased from 0.417, 1.521, and 2.755 mA to 0.414, 1.483, and 2.707 mA, which is a reduction of 0.7%, 2.5%, and 1.7%, respectively. As the LET increases, the total amount of peak pulse increases. To distinguish the contribution of bipolar amplification current and drift current to the current peak, the peak pulse current corresponding to the diode is divided by the peak pulse current of the overall device, as shown in Figure 8, with a temperature point of 296 K selected. As the LET value increases, the contribution of the drift current and the bipolar amplification current to the peak pulse current of the device shows different trends. When the LET value increases from 1 to 10 Mev·cm^2/mg, the proportion of drift current increases from 50.2% to 62.3%, while the proportion of bipolar amplification current decreases from 49.8% to 37.7%.

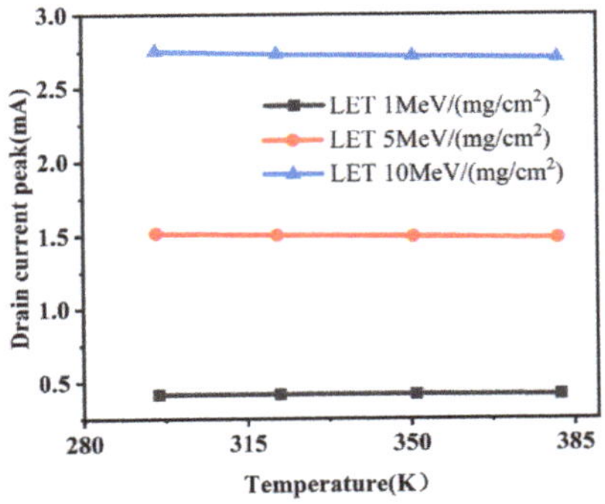

Figure 7. Temperature dependence of peak pulse current.

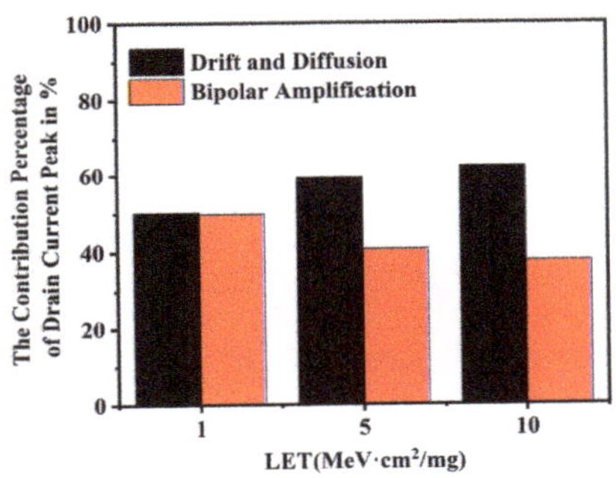

Figure 8. Contribution of drift and bipolar amplification to pulse current peak (T = 296 K, Red: drift diffusion current, Black: bipolar amplification current).

Figure 9 shows the variation of the corresponding peak pulse current of the diode with temperature, which decreased from 0.209, 0.903, and 1.717 mA to 0.194, 0.799, and 1.473 mA for LET of 1, 5, and 10 Mev·cm^2/mg when the temperature was increased from 296 K to 382 K. This is a reduction of 7.1%, 11.5%, and 14.2%, respectively. The impact of temperature on the peak current of the pulse is more significant when only drift current is considered. An explanation will be given below.

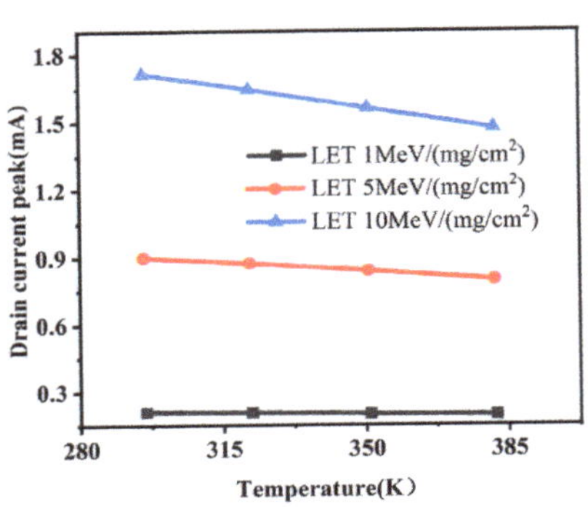

Figure 9. Temperature dependence of drift pulse current.

The amount of charge collected by the drift is related to the carrier mobility, and the relationship between mobility and temperature can be given by the following equation:

$$\mu \propto T^{-3/2} \tag{4}$$

From the equation, it can be seen that mobility is inversely related to temperature. As the lattice vibration becomes more intense at higher temperatures, the probability of carriers receiving scattering increases, resulting in decreased mobility. As shown in Figure 10, the electron mobility around the drain decreases from 525.8 $cm^2/(V{\cdot}s)$ to 313 $cm^2/(V{\cdot}s)$ when the temperature increases from 296 K to 382 K. Thus, the amount of charge collected through the drift mechanism is consequently reduced, so the peak pulse current tends to decrease with increasing temperature for the overall device and the diode. As mentioned earlier, an increase in temperature will cause the bandgap of silicon to decrease, resulting in more electron-hole pairs being generated for the same amount of deposited energy. On the other hand, it will also cause the high energy hot carriers to create more holes through collisional ionization, and these extra holes will further elevate the potential after entering the well region, resulting in the bipolar amplification to collect more charges, which leads to a more significant bipolar amplification effect. Therefore, without considering the bipolar amplification effect, the peak pulse current is more significantly affected by temperature.

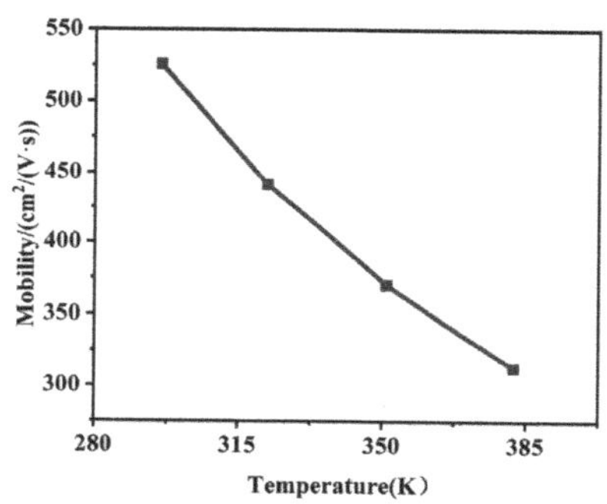

Figure 10. Temperature dependence of electron mobility.

The smaller the pulse peak current, the larger the device's LET threshold, resulting in a reduction in the upset cross-section of the device. From the analysis above, it can be inferred that an increase in temperature reduces drift current while increasing bipolar amplification current. A smaller drift current leads to a larger LET threshold, while a larger bipolar amplification current results in a smaller LET threshold, indicating a competitive relationship between the drift current and bipolar amplification current. From the simulation results, an increase in temperature raises the device's LET threshold, thereby reducing the device's upset cross-section.

3.2.2. Temperature Dependence of Pulse Current Duration

In this paper, the duration is defined as the time from the beginning of the pulse current until it first reaches 0 A. The relationship between pulse current duration and temperature is shown in Figure 11. The LET of the particles is 1 Mev·cm^2/mg. When the temperature increased from 296 K to 382 K, the pulse current duration increased from 13 ps to 17.7 ps, an increase of 36%. The duration of the current is mainly determined by the diffusion mechanism and the bipolar amplification mechanism. As discussed previously, the bipolar amplification current increases with increasing temperature, which leads to an increase in duration. The variation of the diffusion coefficient with temperature can be derived from the following equation [26]:

$$D_{iff} = \frac{\mu KT}{q} \quad (5)$$

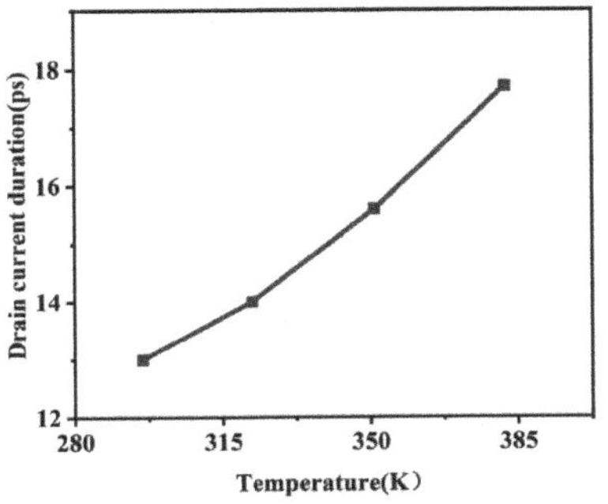

Figure 11. Temperature dependence of pulse current duration.

When the temperature increases, the mobility and the diffusion coefficient decrease, which leads to an increase in the current duration. However, for SRAM, the current duration is up to tens of picoseconds, much longer than the upset time, so this duration will not be considered for SEU.

3.3. Temperature Dependence of Electrical Characteristics

Temperature changes will cause changes in the electrical characteristics of the device. When the particle is incident to the device, it will produce a transient voltage. If the transient voltage exceeds a specific value, it will cause the data to upset, called upset voltage. The upset voltage decreases with increasing temperature, while a decrease in the upset voltage decreases the LET threshold value of the device, which increases the SEU cross-section of the SRAM. The empirical relationship between upset voltage and temperature is as follows [10]:

$$V_{upset}(T) = KT + V_{upset0} \quad (6)$$

T is the operating temperature of the device, K = dV/dT; V_{upset0} is the upset voltage at 0 K, which is determined by the electrical characteristics of the device, and the linear relationship between the upset voltage and the temperature is -0.2 mV/k [10]. The critical charge can be estimated by the following equation:

$$Q_c = C_g V_{upset} \quad (7)$$

Q_c is the critical charge, C_g is the gate capacitance, and the gate capacitance of the 28 nm transistor is generally 0.5 fF. According to Equation (7), it can be concluded that the relationship of the critical charge with the temperature is -1×10^{-3} fC/k. When the temperature rises from 296 K to 382 K, the critical charge decreases by 0.0086 fC. The relationship between critical charge and temperature is approximately linear, and the decrease in critical charge is the main reason for the increase in cross-section. Therefore, the relationship between temperature and cross-section variation also appears to be approximately linear.

Literature [27] based on a combination of large amounts of modeling and experimental data suggests that the critical charge and the characteristic dimensions of the transistor are squarely related:

$$Q_c = 0.23L^2 \quad (8)$$

Q_c is the critical charge, and L is the feature size of the device. According to the above equation, the critical charge of 28 nm SRAM is 0.18 fC. When the temperature increased from 296 K to 382 K, the critical charge decreased by 4.8%. The critical charge of devices produced by different manufacturers will differ for the same feature size. Literature [28] gives that the critical charge of the device is about 0.8 fC for the 65 nm feature size and 0.1 fC for the 22 nm feature size, which is basically comparable to the above equations. The temperature dependence of the experimental device in this paper is more pronounced compared to the devices in the literature [12] (with a feature size of 180 nm), which is due to the fact that the critical charge decreases in square steps as the feature size decreases, and thus also leads to an increasingly severe effect of temperature on the critical charge, which is the main reason for the increase in the upset cross-section. As the feature size continues to decrease, the impact on the SEU cross-section of the decrease in critical charge from changing temperature will become more severe.

4. Conclusions

Based on the ANIS, the temperature dependence of the SEU cross-section in 28 nm embedded SRAM of SiP in the temperature range of 296–382 K was investigated. Under atmospheric neutron irradiation, the SEU cross-section increased from 1.05×10^{-14} cm^2/bit to 1.468×10^{-14} cm^2/bit, an increase of 39.8%, when the temperature was increased from 296 K to 382 K.

The variation of SEU cross-section with temperature is mainly caused by the combination of the variation of the electrical characteristics and the transient pulse current with temperature. The decrease of the peak pulse current increases the LET threshold, while the decrease of the critical charge decreases the LET threshold. Raising the LET threshold will decrease the upset cross-section, while lowering the LET threshold will increase the upset cross-section, both of which play a competitive relationship. TCAD simulation results show that when the temperature was increased from 296 K to 382 K, the transient pulse current of the device decreased by 0.7%, 2.5%, and 1.7% under the irradiation of particles with LET of 1, 5, and 10 Mev·cm^2/mg, respectively. The temperature does not significantly impact it, which is mainly caused by the fact that the bipolar amplification effect becomes more significant with increasing temperature. By increasing the body contact, the bipolar amplification effect on the peak value of the transient pulse current can be reduced, which can enhance the upset resistance of the device and its reliability. The pulse current duration up to tens of picoseconds, which is much longer than the SEU time, is not considered. As the temperature increases from 296 K to 382 K, the critical charge of the device decreases by about 4.8%. As the feature size advances further, the impact of changing temperature on the critical charge will become more significant. The relationship between critical charge and temperature is linear, and the decrease in critical charge is the main reason for the growth of the upset cross-section. Therefore, the relationship between upset cross-section and temperature is approximately linear. The above conclusions provide a theoretical basis for the widespread use of SRAM in aerospace environments at temperatures of 296–382 K and for future development of advanced SRAM.

Author Contributions: Conceptualization, S.Z.; Methodology, S.Z.; Software, S.Z., X.L. and H.Z.; Validation, Z.L. (Zhili Liu), G.G. and Q.Z.; Formal analysis, S.Z.; Investigation, S.Z.; Data curation, S.Z. and J.Y.; Writing—original draft, S.Z.; Writing—review & editing, Z.Z.; Supervision, Z.Z., Z.L. (Zhili Liu), G.G., Q.Z. and H.L.; Project administration, Z.Z., Z.L. (Zhifeng Lei), Z.L. (Zhili Liu), G.G. and Q.Z.; Funding acquisition, Z.Z. and Z.L. (Zhifeng Lei). All authors have read and agreed to the published version of the manuscript.

Funding: This work was supported by the National Natural Science Foundation of China (Grant Nos. 12175045, 12075065), the Key-Area Research and Development Program of Guangdong Province (Grant No. 2022B0701180002) and the Opening Project of Science and Technology on Reliability Physics and Application Technology of Electronic Component Laboratory.

Data Availability Statement: Data are contained within the article.

Acknowledgments: The authors sincerely thank colleagues of China Spallation Neutron Source for their great support and assistance during the experiment.

Conflicts of Interest: Authors Zhili Liu, Gaoying Geng and Qi Zhang were employed by the company The 58th Institute of China Electronics Technology Group Corporation. The remaining authors declare that the research was conducted in the absence of any commercial or financial relationships that could be construed as a potential conflict of interest.

References

1. Abe, S.; Watanabe, Y. Analysis of Charge Deposition and Collection Caused by Low Energy Neutrons in a 25-Nm Bulk CMOS Technology. *IEEE Trans. Nucl. Sci.* **2014**, *61*, 3519–3526. [CrossRef]
2. Hubert, G.; Bezerra, F.; Nicot, J.-M.; Artola, L.; Cheminet, A.; Valdivia, J.-N.; Mouret, J.-M.; Meyer, J.-R.; Cocquerez, P. Atmospheric Radiation Environment Effects on Electronic Balloon Board Observed During Polar Vortex and Equatorial Operational Campaigns. *IEEE Trans. Nucl. Sci.* **2014**, *61*, 1703–1709. [CrossRef]
3. Dyer, C.; Hands, A.; Ryden, K.; Lei, F. Extreme Atmospheric Radiation Environments and Single Event Effects. *IEEE Trans. Nucl. Sci.* **2018**, *65*, 432–438. [CrossRef]
4. Normand, E. Single Event Upset at Ground Level. *IEEE Trans. Nucl. Sci.* **1996**, *43*, 2742–2750. [CrossRef]
5. Normand, E. Single-Event Effects in Avionics. *IEEE Trans. Nucl. Sci.* **1996**, *43*, 461–474. [CrossRef]
6. Gadlage, M.J.; Ahlbin, J.R.; Ramachandran, V.; Gouker, P.; Dinkins, C.A.; Bhuva, B.L.; Narasimham, B.; Schrimpf, R.D.; McCurdy, M.W.; Alles, M.L.; et al. Temperature Dependence of Digital Single-Event Transients in Bulk and Fully-Depleted SOI Technologies. *IEEE Trans. Nucl. Sci.* **2009**, *56*, 3115–3121. [CrossRef]
7. Laird, J.S.; Hirao, T.; Onoda, S.; Mori, H.; Itoh, H. Temperature Dependence of Heavy Ion-Induced Current Transients in Si Epilayer Devices. *IEEE Trans. Nucl. Sci.* **2002**, *49*, 1389–1395. [CrossRef]
8. Guo, G.; Hirao, T.; Laird, J.S.; Onoda, S.; Wakasa, T.; Yamakawa, T.; Kamiya, T. Temperature Dependence of Single-Event Transient Current Induced by Heavy-Ion Microbeam on p/Sup +//n/n/Sup +/ Epilayer Junctions. *IEEE Trans. Nucl. Sci.* **2004**, *51*, 2834–2839. [CrossRef]
9. Liu, T.; Liu, J.; Geng, C.; Zhang, Z.; Zhao, F.; Tong, T.; Sun, Y.; Su, H.; Yao, H.; Gu, S.; et al. Influence of Deposited Energy in Sensitive Volume on Temperature Dependence of SEU Sensitivity in SRAM Devices. In Proceedings of the 2013 14th European Conference on Radiation and Its Effects on Components and Systems (RADECS), Oxford, UK, 23–27 September 2013; IEEE: Oxford, UK, 2013; pp. 1–6.
10. Cai, L.; Guo, G.; Liu, J.-C.; Fan, H.; Shi, S.-T.; Wang, H.; Wang, G.-L.; Shen, D.-J.; Hui, N.; He, A.-L. Experimental Study of Temperature Dependence of Single-Event Upset in SRAMs. *Nucl. Sci. Tech.* **2016**, *27*, 16. [CrossRef]
11. Liu, B.; Chen, S.; Liang, B.; Liu, Z.; Zhao, Z. Temperature Dependency of Charge Sharing and MBU Sensitivity in 130-nm CMOS Technology. *IEEE Trans. Nucl. Sci.* **2009**, *56*, 2473–2479. [CrossRef]
12. Bagatin, M.; Gerardin, S.; Paccagnella, A.; Andreani, C.; Gorini, G.; Frost, C.D. Temperature Dependence of Neutron-Induced Soft Errors in SRAMs. *Microelectron. Reliab.* **2012**, *52*, 289–293. [CrossRef]
13. Cao, J.; Xu, L.; Wen, S.-J.; Fung, R.; Narasimham, B.; Massengill, L.W.; Bhuva, B.L. Temperature Dependence of Single-Event Transient Pulse Widths for 7-Nm Bulk FinFET Technology. In Proceedings of the 2020 IEEE International Reliability Physics Symposium (IRPS), Dallas, TX, USA, 28 April–30 May 2020; pp. 1–5.
14. Cai, L.; Chi, Y.-Q.; Ye, B.; Liu, Y.-Z.; He, Z.; Wang, H.-B.; Sun, Q.; Sun, R.-Q.; Gao, S.; Hu, P.-P.; et al. Effect of Temperature on Heavy Ion-Induced Single Event Transient on 16-Nm FinFET Inverter Chains. *Chin. Phys. B* **2023**, *32*, 046101. [CrossRef]
15. Olsen, J.; Becher, P.E.; Fynbo, P.B.; Raaby, P.; Schultz, J. Neutron-Induced Single Event Upsets in Static RAMS Observed a 10 km Flight Attitude. *IEEE Trans. Nucl. Sci.* **1993**, *40*, 74–77. [CrossRef]
16. Taber, A.; Normand, E. Single Event Upset in Avionics. *IEEE Trans. Nucl. Sci.* **1993**, *40*, 120–126. [CrossRef]
17. Zhang, Z.-G.; Lei, Z.-F.; Tong, T.; Li, X.-H.; Wang, S.-L.; Liang, T.-J.; Xi, K.; Peng, C.; He, Y.-J.; Huang, Y.; et al. Comparison of Neutron Induced Single Event Upsets in 14 Nm FinFET and 65 Nm Planar Static Random Access Memory Devices. *Acta Phys. Sin.* **2020**, *69*, 056101. [CrossRef]
18. Yang, W.; Li, Y.; Li, Y.; Hu, Z.; Xie, F.; He, C.; Wang, S.; Zhou, B.; He, H.; Khan, W.; et al. Atmospheric Neutron Single Event Effect Test on Xilinx 28 Nm System on Chip at CSNS-BL09. *Microelectron. Reliab.* **2019**, *99*, 119–124. [CrossRef]
19. Yu, Q.; Shen, F.; Yuan, L.; Lin, L.; Hu, Z.; Zhou, B.; Liang, T. Physical Design of an Atmospheric Neutron Irradiation Spectrometer at China Spallation Neutron Source. *Nucl. Eng. Des.* **2022**, *386*, 111579. [CrossRef]
20. *JESD89A*; Measurement and Reporting of Alpha Particle and Terrestrial Cosmic Ray-Induced Soft Errors in Semiconductor Devices. JEDEC: Arlington County, VA, USA, 2006.
21. Kauppila, J.S.; Kay, W.H.; Haeffner, T.D.; Rauch, D.L.; Assis, T.R.; Mahatme, N.N.; Gaspard, N.J.; Bhuva, B.L.; Alles, M.L.; Holman, W.T.; et al. Single-Event Upset Characterization Across Temperature and Supply Voltage for a 20-Nm Bulk Planar CMOS Technology. *IEEE Trans. Nucl. Sci.* **2015**, *62*, 2613–2619. [CrossRef]
22. Amusan, O.A.; Witulski, A.F.; Massengill, L.W.; Bhuva, B.L.; Fleming, P.R.; Alles, M.L.; Sternberg, A.L.; Black, J.D.; Schrimpf, R.D. Charge Collection and Charge Sharing in a 130 Nm CMOS Technology. *IEEE Trans. Nucl. Sci.* **2006**, *53*, 3253–3258. [CrossRef]

23. Shuming, C.; Bin, L.; Biwei, L.; Zheng, L. Temperature Dependence of Digital SET Pulse Width in Bulk and SOI Technologies. *IEEE Trans. Nucl. Sci.* **2008**, *55*, 2914–2920. [CrossRef]
24. Emery, F.E.; Rabson, T.A. Average Energy Expended Per Ionized Electron-Hole Pair in Silicon and Germanium as a Function of Temperature. *Phys. Rev.* **1965**, *140*, A2089–A2093. [CrossRef]
25. Xu, J.; Guo, Y.; Song, R.; Liang, B.; Chi, Y. Supply Voltage and Temperature Dependence of Single-Event Transient in 28-nm FDSOI MOSFETs. *Symmetry* **2019**, *11*, 793. [CrossRef]
26. Truyen, D.; Boch, J.; Sagnes, B.; Renaud, N.; Leduc, E.; Arnal, S.; Saigne, F. Temperature Effect on Heavy-Ion Induced Parasitic Current on SRAM by Device Simulation: Effect on SEU Sensitivity. *IEEE Trans. Nucl. Sci.* **2007**, *54*, 1025–1029. [CrossRef]
27. Warren, K.M.; Weller, R.A.; Mendenhall, M.H.; Reed, R.A.; Ball, D.R.; Howe, C.L.; Olson, B.D.; Alles, M.L.; Massengill, L.W.; Schrimpf, R.D.; et al. The Contribution of Nuclear Reactions to Heavy Ion Single Event Upset Cross-Section Measurements in a High-Density SEU Hardened SRAM. *IEEE Trans. Nucl. Sci.* **2005**, *52*, 2125–2131. [CrossRef]
28. Ibe, E.; Taniguchi, H.; Yahagi, Y.; Shimbo, K.; Toba, T. Impact of Scaling on Neutron-Induced Soft Error in SRAMs From a 250 Nm to a 22 Nm Design Rule. *IEEE Trans. Electron. Devices* **2010**, *57*, 1527–1538. [CrossRef]

Disclaimer/Publisher's Note: The statements, opinions and data contained in all publications are solely those of the individual author(s) and contributor(s) and not of MDPI and/or the editor(s). MDPI and/or the editor(s) disclaim responsibility for any injury to people or property resulting from any ideas, methods, instructions or products referred to in the content.

electronics

MDPI

Article

Temperature Dependence of Total Ionizing Dose Effects of β-Ga_2O_3 Schottky Barrier Diodes

Weili Fu [1,2], Teng Ma [1,*], Zhifeng Lei [1], Chao Peng [1], Hong Zhang [1], Zhangang Zhang [1], Tao Xiao [1,2], Hongjia Song [2], Yuangang Wang [3], Jinbin Wang [2], Zhao Fu [2] and Xiangli Zhong [2,*]

1 Reliability Physics and Application Technology of Electronic Component Key Laboratory, China Electronic Product Reliability and Environmental Testing Research Institute, Guangzhou 511370, China
2 National-Provincial Laboratory of Special Function Thin Film Materials, School of Materials Science and Engineering, Xiangtan University, Xiangtan 411105, China
3 The National Key Laboratory of Solid-State Microwave Devices and Circuits, Hebei Semiconductor Research Institute, Shijiazhuang 050051, China
* Correspondence: mateng@ceprei.com (T.M.); xlzhong@xtu.edu.cn (X.Z.)

Abstract: This paper investigates the temperature-dependent effects of gamma-ray irradiation on β-Ga_2O_3 vertical Schottky barrier diodes (SBDs) under a 100 V reverse bias condition at a total dose of 1 Mrad(Si). As the irradiation dose increased, the radiation damage became more severe. The total ionizing dose (TID) degradation behavior and mechanisms were evaluated through DC, capacitance–voltage (C-V), and low-frequency noise (LFN) measurements by varying irradiation, and the test results indicated that TID effects introduced interface defects and altered the carrier concentration within the material. The impact of TID effects was more pronounced at lower temperatures compared to higher temperatures. Additionally, the annealing effect in the high-temperature experimental conditions ameliorated the growth of interface trap defects caused by irradiation. These results suggest that compared to low-temperature testing, the device exhibits higher TID tolerance after high-temperature exposure, providing valuable insights for in-depth radiation reliability studies on subsequent related devices.

Keywords: β-Ga_2O_3 Schottky barrier diode (SBD); temperature dependence; gamma-ray irradiation; interface defects; total ionizing dose (TID)

1. Introduction

Compared to traditional narrow-bandgap semiconductor materials like Si and GaAs, as well as other wide-bandgap semiconductor materials such as GaN and SiC, the ultra-wide-bandgap semiconductor gallium oxide exhibits exceptional characteristics, including a larger bandgap width, a stronger breakdown electric field, high-temperature resistance, and radiation resistance [1–5]. These features make it a promising material for electronic devices, particularly in high-radiation environments.

However, it is crucial to note that these electronic devices are susceptible to particle-induced deterioration or malfunction [6]. Statistics reveal that approximately 40% of satellite failures are attributed to anomalies caused by radiation effects in space. Consequently, investigating the damage mechanism of spacecraft electronic devices in the space radiation environment and establishing ground simulation test methods for space radiation effects are essential, which enable the prediction of damage caused by the space radiation environment on the ground, thereby improving the operational life and reliability of spacecraft in orbit.

For semiconductor devices made of wide-bandgap materials such as GaN and SiC, total ionizing dose (TID) radiation experiments on GaN-based diodes were conducted in 2020 by Bian et al. In their study, the diodes were subjected to cumulative doses of 1 Mrad(Si) gamma-ray irradiation at room temperature and increasing forward current density [7]. In 2002, S. Metzger et al., reported on the linearity of the response of a SiC ultraviolet photodiode detector after a total dose of 20 kGy (air); the experiments were conducted at five constant temperatures ranging from 23 °C to 200 °C, and no significant changes were observed after testing [8]. In 2019, Pincuda et al., reported on the influence of high-temperature annealing on the threshold voltage drift in SiC MOSFETs after TID effects [9].

While recent studies have reported on the effects of protons, heavy ions, electrons, and neutron irradiation on β-Ga_2O_3 material and device properties, there is limited literature reporting on the impact of the temperature of gamma-ray irradiation on β-Ga_2O_3 SBD [10–12]. Chang et al., reported the impact of 100 MeV high-energy protons on β-Ga_2O_3-based solar-blind photodetectors, finding that a considerable number of oxygen vacancies were formed at the metal–semiconductor interface after irradiation. In 2023, R. M. Cadena et al., reported on low-energy ion-induced breakdown and single-event burnout (SEB) in β-Ga_2O_3 Schottky diodes; the study revealed fundamentally different responses among alpha particles, Cf-252, and heavy-ion irradiation. For power devices like gallium oxide diodes, Liu et al., reported the radiation effect of gamma irradiation (^{60}Co) on Au/Ni/β-Ga_2O_3 vertical Schottky barrier diodes; in addition, the carrier concentration calculated from the C–V measurements increased slightly after gamma irradiation. These results suggest that β-Ga_2O_3 SBDs have high intrinsic gamma irradiation hardness, and operating under bias voltage is inevitable. Due to the less-than-ideal thermal conductivity of gallium oxide material, studying the effects of TID under different temperature levels is crucial.

In this study, we performed temperature-controlled TID radiation experiments on β-Ga_2O_3 SBDs at low-to-high temperatures (−25 °C, 0 °C, 25 °C (room temperature), 50 °C, 75 °C, and 100 °C) using liquid nitrogen refrigeration equipment and a temperature-regulated incubator at 100 V, with a cumulative dose of 1 Mrad(Si), used to meet the test needs of temperature changes during the operation of spacecraft in orbit; the results indicate that the forward current density of the devices not only increases with the irradiation dose but also rises with the temperatures. Through analysis methods such as capacitance–voltage (C-V) testing and low-frequency noise (LFN), the impact of interface defects on device performance is quite pronounced, and high-temperature annealing provides a new approach to reducing interface trap concentration, improving the performance of gallium oxide-based devices.

2. Materials and Methods

The fabrication process of the devices unfolded as follows: Following the formation of N^- type material (net doping concentration of 3.0×10^{16} cm^{-3}, silicon-doped), the epitaxial film was grown using halide vapor-phase epitaxy (HVPE) on an N^+ type (001) crystal-phase β-Ga_2O_3 substrate (net doping concentration of 1.0×10^{19} cm^{-3}, stannum-doped), and surface defects arising during the vapor-phase epitaxial growth of the halide layer were eliminated through chemical mechanical grinding (CMP) technology. The schematic cross-sectional structure of the Schottky barrier diode (SBD) is depicted in Figure 1.

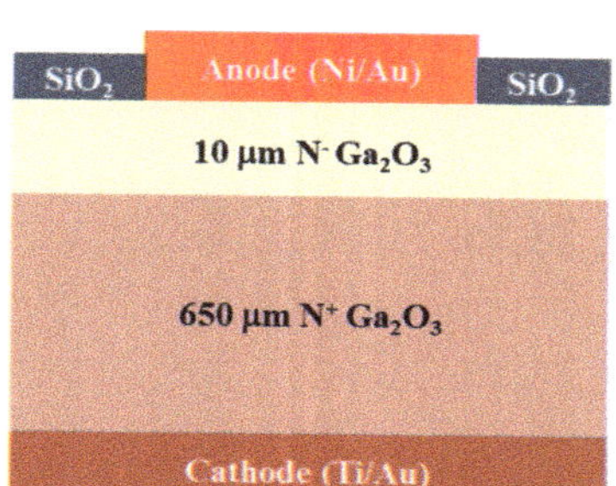

Figure 1. Schematic cross-section of β-Ga_2O_3 SBD structure.

Subsequently, the damage caused by the backside of the substrate during wafer preparation was polished. BCl_3 was applied to the substrate's backside for reactive ion etching, concluding with the evaporation of the Ti/Au stack structure to establish an ohmic contact, which was deposited by electron-beam evaporation, forming the cathode of the devices. The thicknesses were 320 μm and 200 nm. Before device fabrication, the sample was cleaned ultra-sonically several times in acetone, isopropanol, deionized water, and buffered oxide etch, and separately, a thin 200 nm SiO_2 film was grown on the epitaxial layer through plasma-enhanced chemical vapor deposition. A round metal anode was evaporated and stripped by Ni/Au evaporation in a SiO_2 window etched using a photolithography-patterned and buffered hydrofluoric acid solution. The thicknesses were 50 nm and 200 nm, and prior to the fabrication of the anode, the device was thermally oxidized in an oxygen atmosphere at a temperature of 400 °C for 20 min, which reduced the concentration of electrons in the anode edge region of the device, effectively mitigating the peak field at the edge of the device, thereby increasing the reverse breakdown voltage of the device.

For this SBD, the impact of TID effects on the electrical performance of the devices was studied at different temperatures under a reverse bias condition of 100 V, involving the following specific experimental process. The experiments were conducted at the ^{60}Co source of Xinjiang Technical Institute of Physics and Chemistry Chinese Academy of Sciences, China. Gamma-ray irradiation experiments were carried out at temperatures of −25 °C, 0 °C, 25 °C (room temperature), 50 °C, 75 °C, and 100 °C with a reverse bias condition of 100 V, and the radiation dose rate was 50 rad(Si)/s. The electrical performance, C-V, and LFN of the SBD were tested using the B1500A semiconductor parameter analyzer from Keysight Technologies and the Fs-Pro testing instrument from Primarius Electronics. Measurements were performed under all experimental conditions, at least three devices of each type and size were evaluated, and test junctions are averages. The tests were conducted offline at cumulative doses of 0 krad(Si), 300 krad(Si), 500 krad(Si), and 1 Mrad(Si).

3. Results and Discussions

In Figure 2a, it can be observed that the forward conduction current of devices increased as the cumulative radiation dose increased when experiments were conducted at 25 °C. In Figure 2b, after gamma-ray irradiation with a cumulative dose of 1 Mrad(Si), devices tested at six different temperature conditions showed varying increases in forward current density at 1 V. For instance, at 25 °C, the forward current density at 1 V increased by 25%, while at 100 °C, it increased by 72%. Therefore, the impact on the forward conduction characteristics of the devices became more pronounced with higher temperatures during the irradiation process. Given the high carrier mobility and the substantial average free path inherent in gallium oxide materials, the predominant current transport mechanism in Schottky barriers is primarily attributed to the thermal electron emission of the majority of carriers; following the theory of hot electron emission, the variation in forward current density with voltage can be expressed as described in [13–15], as follows:

$$J = A^{*}T^{2}\exp\left(-\frac{q\phi}{kT}\right)\exp\left(-\frac{qV}{nkT}\right) = J_{sT}exp\left(-\frac{qV}{nkT}\right) \tag{1}$$

$$J_{ST} = A^{*}T^{2}\exp\left(-\frac{q\phi}{kT}\right) A^{*} = \frac{4\pi q m_n{}^{*}k^{2}}{h^{3}} \tag{2}$$

$$\Phi = \frac{kT}{q}\ln\left(\frac{AA^{*}T^{2}}{I_s}\right) \tag{3}$$

$$n = \frac{q}{kT}\frac{\partial V}{\partial(lnJ)} \tag{4}$$

where J is the current density; T is the thermodynamic temperature; $q = 1.6 \times 10^{-19}$ C, which is the electron charge; k is the Boltzmann constant; n represents the Schottky diode and ideal in the experiment, the ideality factor of the Schottky diode; J_{ST} is the reverse saturation current density, which is independent of the applied voltage and is a function of strong temperature dependence; Φ is the Schottky potential, the base height, where A^* is the Richardson constant and is usually taken as 40.8 A/(cm^2·K^2); m_n^* is the effective mass of the electron. We extracted the electrical parameters and the variation in the irradiated diode influenced by the data in Figure 2 and the above formulas, and the changes before and after irradiation with a cumulative dose of 1 Mrad(Si) are presented in Table 1.

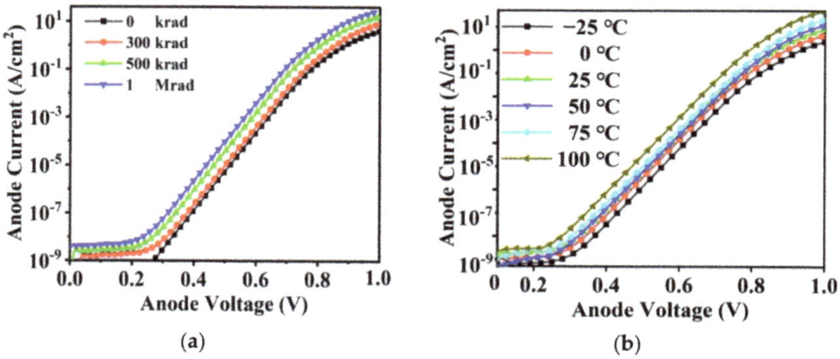

Figure 2. (**a**) Forward current density (J-V) in semi-log-scale at different doses (0 krad(Si), 300 krad(Si), 500 krad(Si), and 1 Mrad(Si)) at a temperature of room temperature and (**b**) forward current density (lgJ-V) in semi-log-scale at different temperature levels (−25 °C, 0 °C, 25 °C, 50 °C, 75 °C, and 100 °C) with a cumulative dose of 1 Mrad(Si).

Table 1. Changes in device electrical parameters before and after irradiation.

	n (before)	n (after)	n	Φ (before)	Φ (after)	Φ	J
−25 °C	1.166	1.147	−1.63%	1.06	1.01	−4.3%	+14%
0 °C	1.19	1.161	−2.44%	0.987	0.951	−3.6%	+19%
25 °C	1.172	1.153	−1.6%	1.032	0.974	−5.6%	+25%
50 °C	1.168	1.156	−1.02%	1.031	0.977	−5.2%	+44%
75 °C	1.156	1.143	−1.12%	1.102	1.057	−4.1%	+57%
100 °C	1.174	1.140	−2.90%	1.024	0.985	−3.81%	+72%

From Table 1, we can see that except for the increase in forward current density with the experimental temperature, the changes in φ and n were not significant. As the temperature increased, more electrons within the gallium oxide material gained energy, transitioning from the valence band to the conduction band to participate in conduction; simultaneously, the ionizing damage caused by irradiation increased with the accumulation of irradiation dose. These two factors contributed to the systematic changes in the forward current density of the devices. The changes in the reverse electrical characteristics of the devices are shown in Figure 3; in Figure 3a, it is shown that the reverse current density of the devices increased with the cumulative radiation dose. For this SBD, when the cathode voltage exceeded a certain critical value, the device current suddenly increased, and this reverse voltage was defined as the breakdown voltage of the diode; before the irradiation experiment, the breakdown voltage of the device was 650 V. As shown in Figure 3b, after irradiation at different temperatures, the breakdown voltage of the devices decreased; compared to that at lower temperatures, the reduction in breakdown voltage was relatively lower at higher temperatures.

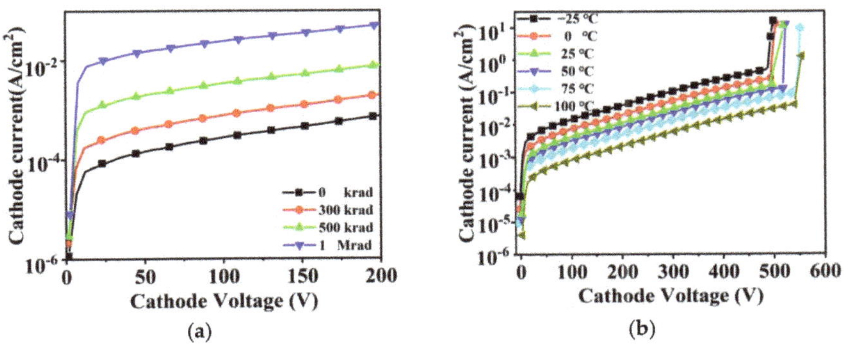

Figure 3. (**a**) Reverse current density (J-V) in semi-log-scale after different doses of gamma-ray radiation at room temperature; (**b**) reverse current density (J-V) in the semi-log-scale plot at different temperatures (−25 °C, 0 °C, 25 °C, 50 °C, 75 °C, and 100 °C) after 1 Mrad(Si) doses of gamma-ray radiation.

Reverse leakage current is a crucial parameter influencing breakdown voltage. In the SBDs, the reverse leakage current of the device increased with the accumulation of irradiation dose before reaching the reverse voltage required for breakdown. The reverse leakage current in the SBDs mainly consisted of thermal electron emission current and recombination current; the former is strongly correlated with the Schottky barrier, while the latter is influenced by carrier lifetime. From Table 1, it can be inferred that gamma-ray irradiation led to a slight reduction in the Schottky barrier height (SBH), increasing thermionic emission current, contributing to the observed slight increase in reverse leakage current and the decrease in breakdown voltage. However, in Figure 3b, it is evident that there was a slight decrease in reverse leakage current density with increasing irradiation temperature. To further understand this phenomenon, we conducted C-V tests, and the results are shown in Figure 4.

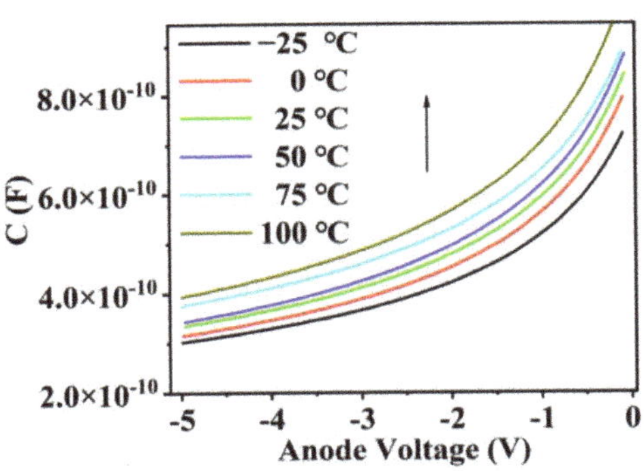

Figure 4. β-Ga_2O_3 SBD C-V plot after 1 Mrad(Si) dose of gamma-ray radiation at different temperatures (−25 °C, 0 °C, 25 °C, 50 °C, 75 °C, and 100 °C) while the frequency is 1 MHz.

The corresponding carrier concentration in the drift layer can be obtained from the following formula [16–18]:

$$N_d = -\frac{2}{q\varepsilon\varepsilon_0 A^2} \times \frac{1}{\frac{dC^{-2}}{dV}} \tag{5}$$

where N_d represents the carrier concentration; from the data in Figure 4, we extracted the specific changes in the carrier concentration of the device before and after irradiation, as shown in Table 2.

Table 2. Carrier concentration before and after irradiation in the drift layer.

	N_d (before Irradiation cm^{-3})	N_d (after Irradiation cm^{-3})	Change Magnitude
−25 °C	8.520×10^{15}	7.496×10^{15}	−12.0%
0 °C	8.416×10^{15}	7.740×10^{15}	−8.02%
25 °C	8.526×10^{15}	8.007×10^{15}	−6.08%
50 °C	8.231×10^{15}	7.875×10^{15}	−4.32%
75 °C	8.318×10^{15}	8.097×10^{15}	−2.65%
100 °C	8.532×10^{15}	8.424×10^{15}	−1.26%

As observed in Table 2, the concentration of carriers involved in conduction decreased by 12%, dropping from $8.520 \times 10^{15}/\text{cm}^3$ to $7.496 \times 10^{15}/\text{cm}^3$ after irradiation with 1 Mrad(Si) at an irradiation temperature of −25 °C. However, in devices irradiated at 100 °C, the concentration of carriers participating in conduction decreased from $8.532 \times 10^{15}/\text{cm}^3$ to $8.424 \times 10^{15}/\text{cm}^3$, showing a mere 1.26% decrease, and the carrier concentration of the device exhibited a gradient change from low temperature to high temperature. For semiconductors, the carrier concentration is mainly influenced by the material's bulk defects and interface defects. When a device is exposed to gamma-ray irradiation, it not only generates ionizing damage, leading to a transient increase in forward current, but also introduces additional hole–electron pairs inside the material, and the holes, with low mobility, migrate under applied bias to the metal–semiconductor contact, forming interface defects. Moreover, when the irradiation dose is sufficiently large, bulk defects are also introduced into gallium oxide material. The combined effects of these two types of defects impact the carrier concentration and the electrical characteristics of devices.

Electrical properties serve as external indicators of the internal structure and current conduction of electronic devices; changes in a device's internal structure can be reflected through alterations in its electrical properties. As per the above analysis, TID effects induced damage to the β-Ga_2O_3 SBDs, leading to alterations in the internal conduction characteristics of the device; these included a decrease in the ideality factor and an elevation in forward current post-turn-on. Although these electrical parameters could characterize the radiation resistance of the β-Ga_2O_3 diode, they were not very sensitive indicators due to the subtle nature of their changes.

Noise, on the other hand, is recognized for its sensitivity in reflecting various potential defects that can lead to device failure; noise analysis provides a rapid and non-destructive means of detection. As the lattice structure, electronic state, or impurity distribution of a device undergoes slow changes over time or due to stress, noise exhibits more significant increases and alterations compared to other structurally sensitive parameters (such as leakage current and ideality factor). The failure of semiconductor devices is often attributed to the presence of some underlying defects generated during the production and operation processes of the device; the noise in electronic devices is quite sensitive to the defects present in the device, and therefore, noise testing is often used to diagnose and predict the reliability of devices. Numerous research results have indicated that $1/f$ noise can serve as a sensitive parameter characterizing electronic components [19–21]. Hence, in the study of space charge regions, oxide-layer defects near the oxide semiconductor interface, and interface states, noise diagnostics prove to be a valuable technique, reflecting the degree of radiation damage more effectively than traditional electrical parameters. The test results for LFN are shown in Figure 5.

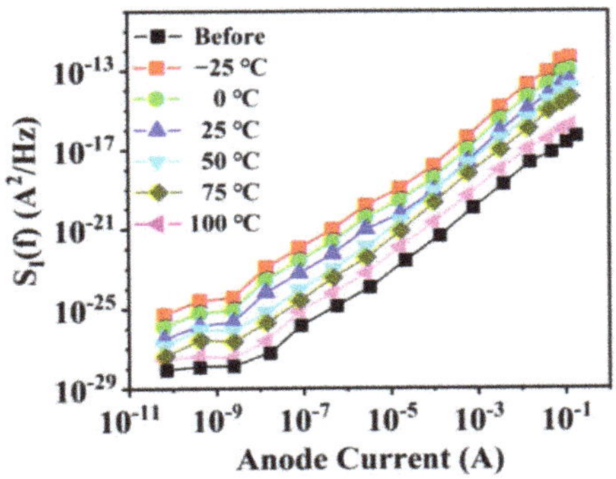

Figure 5. β-Ga_2O_3 SBD LFN plot after 1 Mrad(Si) dose of gamma-ray radiation at different temperatures (−25 °C, 0 °C, 25 °C, 50 °C, 75 °C, and 100 °C) while the frequency is 100 Hz. Before is the device without radiation.

In gallium oxide material, Ga and O atoms have high displacement threshold energies of 20.5 eV and 25.56 eV, respectively. This characteristic gives it higher radiation resistance compared to arsenide materials, and introducing bulk defects in GaAs requires a cumulative dose of gamma ray, reaching 5.45 × 10^{16} rad(Si) [22,23]. Therefore, the total dose of 1 Mrad(Si) in this experiment was not sufficient to introduce bulk defects in Ga_2O_3, a material with stronger radiation resistance; the observed changes in the noise current power spectral density in this experiment excluded the influence of bulk defects. From Figure 5, at lower temperatures, −25 °C and 0 °C, the amplitude of the LFN current power spectral density after irradiation was greater than that at room temperature. However, at higher temperatures, 50 °C, 75 °C, and 100 °C, the amplitude of the LFN current power spectral density was smaller after irradiation compared to at room temperature. Based on the three formulas for LFN in a current diode, we conducted the following analysis to study this mechanism.

Formula (1), HSU's proposed LFN carrier fluctuation model, suggests that the changes in LFN in SBDs result from the modulation of the SBH by traps or fluctuations generated by electron states in composite centers. This is primarily associated with variations in the concentration of interface defects within the material, fundamentally caused by changes in defect concentration, which in turn affect carrier concentration as follows [24]:

$$S_{I_u}=\frac{\beta^2}{f}\left(\frac{qI}{4\varepsilon_s}\right)^2\left(\frac{2h}{4\pi q}\right)^2\frac{qD_t}{6m^*N_dAW^2} \tag{6}$$

where β characterizes the carrier transport mechanism of the SBD which is equal to q/kT for thermionic emission, f is the frequency of the noise test, q is the amount of charge, I is the forward voltage at the time of the test, ε_s is the dielectric constant of the semiconductor material (gallium oxide in this test), h is Planck's constant, D_t is the defect density of the device body, m^* is the effective mass of electrons in the semiconductor material, N_d is the doping concentration after irradiation, A is the effective contact area of the device, and W is the width of the space charge region. Due to the small variation in SBH and as analyzed earlier, the influence of bulk defects is not considered in this experiment. Therefore, Equation (6) does not apply to the current study.

Regarding Formula (2), Luo et al., posit that the 1/f noise observed in SBDs is induced by fluctuations in carrier mobility within the space charge region. They propose models based on the fluctuations in mobility and diffusion coefficient, outlined as follows [25,26]:

$$S_{I_m}=\frac{\alpha I}{16\pi f}\left(\frac{1}{\beta(V_D-V_F)}\right)^{5/2}\cdot\left(\frac{q}{\mu m^*}\right)^2\left(\frac{m^*\varepsilon_S}{\pi N_d}\right)^{1/2} \tag{7}$$

where μ is the electron mobility, α is carrier mobility, V_D is the built-in potential difference of the pn junction, and V_F is the forward bias voltage. From Equation (7), it can be seen that the main variable that affected the noise power spectral density in this study was the change in carrier concentration. Combined with the data from Figures 4 and 5, our preliminary analysis indicates that the change in carrier concentration inside the devices is more pronounced at low temperatures, while the decrease in carrier concentration is smaller at high temperatures, with less fluctuation.

Regarding Formula (3), the concept of the random motion of electrons in Schottky-touched interface states was initially introduced by Jantsch. According to his argument, the interfacial current effect is directly linked to the generation–recombination of interface states. The current–noise density spectrum of the interface state can be expressed as proposed by Jantsch and others [27]. The expression for the current–noise density spectrum of the interface state, as proposed by Jantsch and others, is given by the following:

$$S_{I_r} = \frac{G\beta}{f}\left(\frac{qI}{4\varepsilon_S}\right)^2 \frac{qD_{is}}{\pi N_d WA} \tag{8}$$

where G is a constant with a value of 0.1; D_{is} is the interface state density; $q = 1.6 \times 10^{-19}$ C, which is the electron charge; k is the Boltzmann constant; β characterizes the carrier transport mechanism of the SBD which is equal to q/kT for thermionic emission; f is the frequency of the noise test; q is the amount of charge; I is the forward voltage at the time of the test; ε_S is the dielectric constant of the semiconductor material (gallium oxide in this test); h is Planck's constant; N_d is the doping concentration after irradiation; A is the effective contact area of the device; and W is the width of the space charge region. From this formula, we find that, in addition to N_d, the main variable affecting the noise power spectral density in this study is Dis, which is the interface state defect concentration. At higher temperatures, the annealing effect inside the material leads to a rearrangement of the lattice and other structures at the interface, resulting in a decrease in interface defect concentration. Simultaneously, this reduction leads to a decrease in recombination current, as observed in the changes in carrier concentration and reverse leakage current. The LFN power spectral density of the device is influenced by both carrier concentration and interface defects. Therefore, during high-temperature experiments, the noise power spectral density of the device was relatively smaller, indicating a partial recovery of the device's performance under high-temperature conditions.

Gamma-ray irradiation induces TID effects in devices, leading to transient changes in their electrical properties; it can also generate excess hole–electron pairs in gallium oxide materials and form interface defects in SBDs under the influence of voltage. Once irradiation stops, some defects rapidly return to their original state, leading to the initial steady-state decay of the damage peak, which occurs in approximately 1 s; subsequently, the damage decays to a lower level, and the presence of these defects leads to semi-permanent changes in the device's performance. At higher temperatures, defects in the Ga_2O_3 lattice undergo annealing during the irradiation process; this annealing process allows for the rearrangement and elimination of defects, reducing the defect concentration and restoring the performance parameters of the irradiated device. And without heat treatment, these defects cannot be reduced. Experimental results indicate that under the testing conditions of −25 °C and 0 °C, compared to room temperature, the concentration of interface defects in the irradiated sample is higher. Therefore, the carrier concentration and LFN exhibited larger changes. At 50 °C, 75 °C, and 100 °C, after annealing, the interface defect density was improved, leading to the partial recovery of device performance [28].

4. Conclusions

We conducted gamma-ray irradiation experiments with a cumulative dose of 1 Mrad(Si) under a reverse bias voltage of 100 V across a temperature gradient from −25 °C to 100 °C. The results indicate that the forward conduction current density of the device increases with the cumulative irradiation dose, and the increase is more pronounced at higher temperatures. Because the reverse leakage current decreases with the increase in irradiation temperature, the reverse current of the device exhibits a different trend from the forward current. Analysis through C-V testing and LFN testing indicates that at lower temperatures, the increase in interface trap density caused by irradiation leads to an increase in interface recombination current, resulting in an elevation of leakage current and a decrease in reverse breakdown voltage. This is associated with a reduction in carrier concentration and a decrease in the noise current power spectral density, contributing to the deterioration of device performance. However, compared to room temperature, at higher temperatures, the interface states of the device undergo annealing, allowing for the reordering of interface lattices and a reduction in interface state density. This leads to a decrease in recombination current, a reduction in leakage current, and the recovery of reverse breakdown voltage. The trends in carrier concentration and noise current power spectral density changes also indicate that high-temperature annealing can improve device performance, thereby restoring its reliability.

Author Contributions: Conceptualization, C.P. and Z.Z.; Methodology, H.S. and Z.F.; Validation, H.Z.; Formal analysis, T.X.; Investigation, Z.Z.; Resources, Y.W. and J.W.; Writing—original draft, W.F.; Writing—review & editing, T.M., Z.L. and X.Z. All authors have read and agreed to the published version of the manuscript.

Funding: This research was funded by the fund of the National Natural Science Foundation of China (No. 12305299, 12275230, 12075065, and 12027813), GuangDong Basic and Applied Basic Research Foundation (No. 2022A1515111049), and Innovation Center of Radiation Application (No. KFZC2021020403).

Data Availability Statement: The datasets presented in this article are not readily available because the data are part of an ongoing study.

Acknowledgments: Acknowledgement of financial support from National Natural Science Foundation of China, GuangDong Basic and Applied Basic Research Foundation and Innovation Center of Radiation Application funds.

Conflicts of Interest: The authors declare no conflict of interest.

References

1. Tsao, J.Y.; Chowdhury, S.; Hollis, M.A.; Jena, D.; Johnson, N.M.; Jones, K.A.; Kaplar, R.J.; Rajan, S.; Van de Walle, C.G.; Bellotti, E.; et al. Ultrawide-Bandgap Semiconductors: Research Opportunities and Challenges. *Adv. Electron. Mater.* **2018**, *4*, 1600501. [CrossRef]
2. Kim, M.; Seo, J.-H.; Singisetti, U.; Ma, Z. Recent advances in free-standing single crystalline wide band-gap semiconductors and their applications: GaN, SiC, ZnO, β-Ga_2O_3, and diamond. *J. Mater. Chem. C* **2017**, *5*, 8338–8354. [CrossRef]
3. Galazka, Z. β-Ga_2O_3 for wide-bandgap electronics and optoelectronics. *Semicond. Sci. Technol.* **2018**, *33*, 113001. [CrossRef]
4. Fujita, S. Wide-bandgap semiconductor materials: For their full bloom. *Jpn. J. Appl. Phys.* **2015**, *54*, 030101. [CrossRef]
5. Mastro, M.A.; Kuramata, A.; Calkins, J.; Kim, J.; Ren, F.; Pearton, S.J. Perspective—Opportunities and Future Directions for Ga_2O_3. *ECS J. Solid State Sci. Technol.* **2017**, *6*, P356.
6. Srour, J.R.; McGarrity, J.M. Radiation effects on microelectronics in space. *Proc. IEEE* **1988**, *76*, 1443–1469. [CrossRef]
7. Bian, Z.; Su, K.; Zhang, J.; Zhao, S.; Zhou, H.; Zhang, W.; Zhang, Y.; Zhang, T.; Chen, J.; Dang, K.; et al. Gamma irradiation impact on GaN quasi-vertical Schottky barrier diodes. *J. Phys. D Appl. Phys.* **2020**, *53*, 045103. [CrossRef]
8. Metzger, S.; Henschel, H.; Kohn, O.; Lennartz, W. Silicon carbide radiation detector for harsh environments. *IEEE Trans. Nucl. Sci.* **2002**, *49*, 1351–1355. [CrossRef]
9. Pintacuda, F.; Massetti, S.; Muschitiello, M.; Cantarella, V. TID test results of radiation hardened SiC MOS structures pre-temperature stressed. In Proceedings of the 2019 European Space Power Conference (ESPC), Juan-Les-Pins, France, 30 September–4 October 2019; pp. 1–4.
10. Chang, M.M.; Guo, D.Y.; Zhong, X.L.; Zhang, F.B.; Wang, J.B. Impact of 100 MeV high-energy proton irradiation on β-Ga_2O_3 solar-blind photodetector: Oxygen vacancies formation and resistance switching effect. *J. Appl. Phys.* **2022**, *132*, 123105. [CrossRef]
11. Konishi, K.; Goto, K.; Murakami, H.; Kumagai, Y.; Kuramata, A.; Yamakoshi, S.; Higashiwaki, M. 1-kV vertical Ga_2O_3 field-plated Schottky barrier diodes. *Appl. Phys. Lett.* **2017**, *110*, 103506. [CrossRef]
12. Polyakov, A.Y.; Smirnov, N.B.; Shchemerov, I.V.; Pearton, S.J.; Ren, F.; Chernykh, A.V.; Lagov, P.B.; Kulevoy, T.V. Hole traps and persistent photo capacitance in proton irradiated β-Ga_2O_3 films doped with Si. *APL Mater.* **2018**, *6*, 096102. [CrossRef]

13. Liu, M.; Hua, M.; Tian, X.; Wang, Z.; Gao, H.; Wang, W.; Chen, Y.; Zhang, C.; Zhao, S.; Feng, Q.; et al. Effect of gamma irradiation on β-Ga_2O_3 vertical Schottky barrier diode. *Appl. Phys. Lett.* **2023**, *123*, 212103. [CrossRef]
14. Holmes-Siedle, A.G.; Adams, L. *Handbook of Radiation Effects*; Oxford University Press: Oxford, UK, 1993.
15. Sheoran, H.; Kumar, V.; Singh, R. A Comprehensive Review on Recent Developments in Ohmic and Schottky Contacts on Ga_2O_3 for Device Applications. *ACS Appl. Electron. Mater.* **2022**, *4*, 2589–2628. [CrossRef]
16. Sze, S.M.; Li, Y.; Ng, K.K. *Physics of Semiconductor Devices*; Nework Wiley: Hoboken, NJ, USA, 1981.
17. Zheng, X.-F.; Dong, S.-S.; Ji, P.; Wang, C.; He, Y.-L.; Lv, L.; Ma, X.-H.; Hao, Y. Characterization of bulk traps and interface states in AlGaN/GaN heterostructure under proton irradiation. *Appl. Phys. Lett.* **2018**, *112*, 233504. [CrossRef]
18. Ai, W.-S.; Liu, J.; Feng, Q.; Zhai, P.-F.; Hu, P.-P.; Zeng, J.; Zhang, S.-X.; Li, Z.-Z.; Liu, L.; Yan, X.-Y.; et al. Degradation of β-Ga_2O_3 Schottky barrier diode under swift heavy ion irradiation. *Chin. Phys. B* **2021**, *30*, 056110. [CrossRef]
19. Vandamme, L.K.J. Noise as a diagnostic tool for quality and reliability of electronic devices. *IEEE Trans. Electron Devices* **1994**, *41*, 2176–2187. [CrossRef]
20. Zhuang, Y.; Du, L. 1/f noise as a reliability indicator for subsurface Zener diodes. *Microelectron. Reliab.* **2002**, *42*, 355–360. [CrossRef]
21. Yiqi, Z.; Qing, S. 1f noise as a prediction of long-term instability in integrated operational amplifiers. *Microelectron. Reliab.* **1996**, *36*, 189–193. [CrossRef]
22. Khanna, S.M.; Webb, J.; Tang, H.; Houdayer, A.J.; Carlone, C. 2 MeV proton radiation damage studies of gallium nitride films through low temperature photoluminescence spectroscopy measurements. *IEEE Trans. Nucl. Sci.* **2000**, *47*, 2322–2328. [CrossRef]
23. Khanna, R.; Han, S.Y.; Pearton, S.J.; Schoenfeld, D.W.; Schoenfeld, W.V.; Ren, F.J.A.P.L. High dose Co-60 gamma irradiation of InGaN quantum well light-emitting diodes. *Appl. Phys. Lett.* **2005**, *87*, 212107. [CrossRef]
24. Hsu, T.S. Low-frequency excess noise in metal—Silicon Schottky barrier diodes. *IEEE Trans. Electron Devices* **1970**, *17*, 496–506. [CrossRef]
25. Luo, M.Y.; Bosman, G.; Ziel, A.V.D.; Hench, L.L. Theory and experiments of 1/f noise in Schottky-barrier diodes operating in the thermionic-emission mode. *IEEE Trans. Electron Devices* **1988**, *35*, 1351–1356. [CrossRef]
26. Schiebel, R.A. A model for 1/f noise in diffusion current based on surface recombination velocity fluctuations and insulator trapping. *IEEE Trans. Electron Devices* **1994**, *41*, 768–778. [CrossRef]
27. Jantsch, O. Flicker (1/f) noise generated by a random walk of electrons in interfaces. *IEEE Trans. Electron Devices* **1987**, *34*, 1100–1115. [CrossRef]
28. Zuleeg, R. Radiation effects in GaAs FET devices. *Proc. IEEE* **1989**, *77*, 389–407. [CrossRef]

Disclaimer/Publisher's Note: The statements, opinions and data contained in all publications are solely those of the individual author(s) and contributor(s) and not of MDPI and/or the editor(s). MDPI and/or the editor(s) disclaim responsibility for any injury to people or property resulting from any ideas, methods, instructions or products referred to in the content.

electronics

MDPI

Article

Impact of Air Gaps Between Microstrip Line and Magnetic Sheet on Near-Field Magnetic Shielding

Hyun Ho Park [1,*], Eakhwan Song [2], Jiseong Kim [3] and Cheolsoo Kim [4]

1 School of Electric and Electronic Engineering, The University of Suwon, Hwaseong 18323, Republic of Korea
2 Department of Electronics and Communications Engineering, Kwangwoon University, Seoul 01897, Republic of Korea; esong@kw.ac.kr
3 CCS Graduate School of Mobility, Korea Advanced Institute of Science and Technology, Daejeon 34141, Republic of Korea; js.kim@kaist.ac.kr
4 EMCIS, Co., Ltd., Anyang 14086, Republic of Korea; emiemc@emcis.co.kr
* Correspondence: hhpark@suwon.ac.kr

Abstract: This study experimentally analyzed the impact of air gaps between a magnetic sheet and a test board with a microstrip line, which is used to measure the near-field magnetic shielding effectiveness (NSE) of magnetic sheets made of metallic powder. To conduct the measurements, a material fixture equipped with a microstrip line to generate the near magnetic field, a rectangular loop probe, and an automatic probe positioning system capable of moving the loop probe along three axes were designed and fabricated. In addition, to systematically vary the thickness of the gaps, three paper spacers with a thickness of 0.11 mm per paper were used, and a 1.0 mm thick acrylic sheet, along with a specially designed material fixture, was used to press down the magnetic sheets during measurement. The magnetic shielding properties were measured and compared under various air gap conditions using a near-field magnetic loop probe. The effect of the gaps on the shielding performance of the magnetic sheets was quantitatively evaluated for three different magnetic sheets. The experimental results showed that as the gap thickness increased, NSE tended to improve up to a frequency around 1 GHz, while in the higher frequency range of a few GHz, NSE tended to decrease. The physical background of this phenomenon was explained using an equivalent magnetic circuit represented by reluctances for the structure, where the magnetic sheet is placed above the microstrip line with an air gap. This model helps to elucidate how the presence of the air gap affects the near-field magnetic shielding performance.

Keywords: magnetic sheets; near-field magnetic shielding effectiveness (NSE); equivalent magnetic circuit

1. Introduction

As the speed and integration density of electronic devices continue to increase, the issue of electromagnetic interference (EMI) has become increasingly critical. In high-speed digital circuits, where high-frequency signals are utilized, EMI originating from the near-field region of electromagnetic waves can cause interference between adjacent circuits, degrading the reliability and stability of the entire system. In addition, reducing interference or coupling from external noise or intentional electromagnetic fields is also a critical problem. To address these issues, the importance of near-field EMI shielding techniques is growing, becoming an essential element to minimize circuit-to-circuit interference and ensure system stability. In this context, magnetic sheets are gaining attention as an effective solution for near-field shielding [1–4]. Magnetic sheets, with their high magnetic permeability, can effectively block EMI by absorbing or reflecting near-field magnetic fields [5]. Consequently, they are used in various applications such as high-frequency circuits, antennas, and wireless communication devices. However, the shielding performance of magnetic sheets can vary significantly depending on the installation environment. In particular, the presence of an air gap between the magnetic sheet and the object to be shielded can alter the path of the magnetic field, potentially degrading the shielding effectiveness. Such air gaps are common in actual circuit design and manufacturing processes, and the resulting decrease in shielding performance can negatively impact the overall performance of electronic devices.

To address this issue, this study aims to experimentally analyze the impact of air gaps between the magnetic sheet and the test board on near-field magnetic shielding characteristics. A magnetic sheet was installed on a test board with a 50-ohm microstrip line, and a near-field magnetic loop probe was used to measure the shielding performance under various air gap conditions. The goal of this study is to quantitatively evaluate the effect of the presence and size of air gaps on the shielding performance of magnetic sheets and to propose an appropriate usage of the magnetic sheets in real applications to maximize their shielding efficiency.

The structure of the paper is as follows. Section 2 introduces the near-field magnetic shielding measurement method, and the equipment designed for the measurements, including the microstrip line PCB, a rectangular loop probe, a material fixture, and an automatic probe positioning system (referred to as EME-28G), which functions as a type of scanning system. Section 3 describes the three materials under test (MUTs), the paper spacers intentionally applied to create air gaps, and the acrylic sheet used to press the magnetic sheets. It also presents the setup used to measure the NSE of the MUTs with the fabricated equipment. Section 4 compares the measured NSE for three commercial magnetic sheets based on varying air gaps and discusses the different phenomena in NSE at low and high frequency ranges by using an equivalent magnetic circuit represented by reluctances. Finally, Section 5 offers a brief conclusion.

2. Method and Apparatus for Near-Field Magnetic Shielding Measurement

2.1. Measurement Configuration

In the IEC 62333-2 standard [6], four methods are proposed for measuring and evaluating the electromagnetic noise coupling attenuation characteristics of magnetic sheets. One of these methods, the inter-decoupling ratio measurement, uses two circular loop probes with a diameter of 3 mm to assess the near-field magnetic shielding properties of the magnetic sheet. However, instead of circular loop probes, some studies use a microstrip line to measure shielding effectiveness by generating near-field magnetic fields [7–9]. Utilizing a microstrip line as the source enables the creation of broadband near-field magnetic fields [10]. As shown in Figure 1, this study also employed a test board with a 50 Ω microstrip line, which was used as the near magnetic field source.

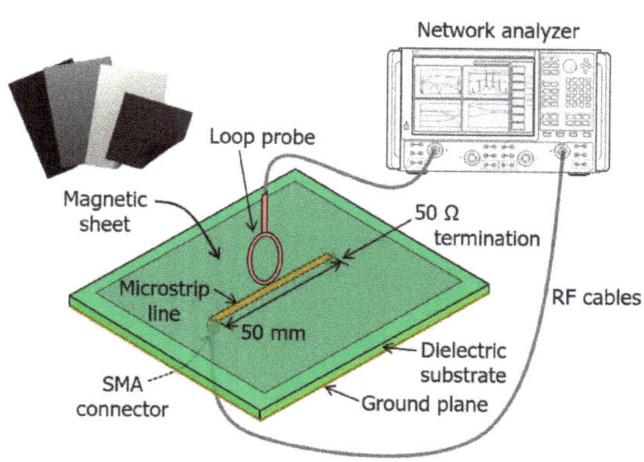

Figure 1. Configuration of near-field magnetic shielding effectiveness (NSE) of magnetic sheets using a 50 Ω microstrip line.

Near-field magnetic shielding effectiveness (NSE) is measured as follows: First, using a vector network analyzer (VNA), the coupling from the microstrip line to the loop probe without the magnetic sheet (S_{21}^{ref}) is measured. Then, with the magnetic sheet in place, the coupling of the near-field magnetic field generated by the microstrip line through the magnetic sheet to the loop probe (S_{21}^{load}) is measured. Finally, the ratio of these measurements is expressed in dB to determine NSE, as given by the following equation.

$$\text{NSE [dB]} = 20log_{10}\left(S_{21}^{ref} / S_{21}^{load}\right),$$

where S_{21}^{ref} and S_{21}^{load} represent the two-port S_{21} parameters between the microstrip line and the VNA when the microstrip board is without and with the magnetic sheet, respectively.

2.2. Design of Measurement Apparatus and Material Fixture

To measure NSE using a microstrip line and loop probe, a 50 Ω microstrip line with a length of 50 mm and a rectangular-shaped loop probe with 2 turns and a loop size of 2 × 3 mm were designed and fabricated, as shown in Figure 2.

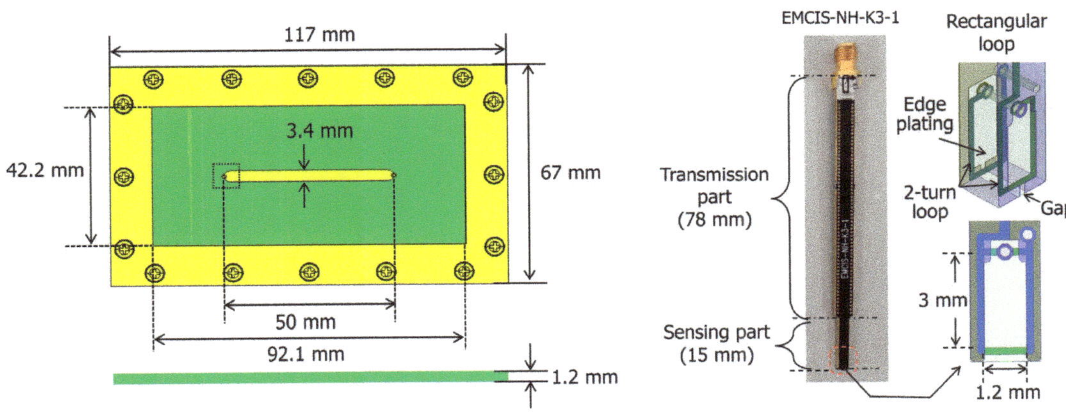

(a) (b)

Figure 2. Design and fabricated 50 Ω microstrip line test board and rectangular loop probe. (**a**) Microstrip line (transmitter). (**b**) Magnetic field probe (receiver).

The microstrip line has a width of 3.4 mm and a length of 50 mm on two-layer printed circuit boards (PCBs). The substrate is made of Teflon with a relative permittivity (ε_r) of approximately 2.5 and a thickness of 1.2 mm. The overall dimensions of the test PCB are 117 mm × 67 mm. The bottom layer of the PCB serves as the ground plane. A rectangular copper ring with a width of 12.4 mm is connected to the ground plane via an edge plating technique around the PCB sides. Additionally, 16 screw holes are drilled into this rectangular ring to establish electrical contact with a material fixture.

To measure the magnetic field generated by the microstrip line, a custom-designed magnetic field probe was utilized. The probe was fabricated using a printed circuit board (PCB) process, offering greater design precision compared to conventional mechanical fabrication methods. The probe features a 4-layer structure: the top and bottom layers provide electromagnetic shielding from external noise, while the inner layers house the sensing and transmission components. The sensing element, designed as a 2-turn rectangular loop, is capable of capturing magnetic fields with a measurement bandwidth of up to 20 GHz. The transmission component transfers the captured magnetic signals to a connector, which interfaces with the measurement equipment. Additionally, the PCB is fully shielded on the sides with edge plating. To further enhance measurement sensitivity, a narrow 0.2 mm slot is incorporated into the bottom of the loop to mitigate eddy currents induced in the shielding structures [11].

Figure 3 shows the coupling property of the designed magnetic loop probe (NH-K3-1) when using the microstrip line as a near magnetic field source, as shown in Figure 2a. The distance between the microstrip line and the bottom of the probe was set to 3 mm. This is the same setup for the reference measurement without a magnetic sheet. To improve loop sensitivity, an amplifier with a 30 dB gain was used during the measurements. To extract the pure coupling efficiency (S21 or Z21) of the loop probe itself, the amplifier gain was subtracted from the measured data in the dB scale. The coupling efficiency (S21 or Z21) is low in the low frequency range but increases linearly by a slop of +20 dB/decade as the frequency increases because the probe acts as an inductance. In the GHz range, the coupling saturates at a constant value, with some resonances caused by the parasitic capacitance of the loop probe.

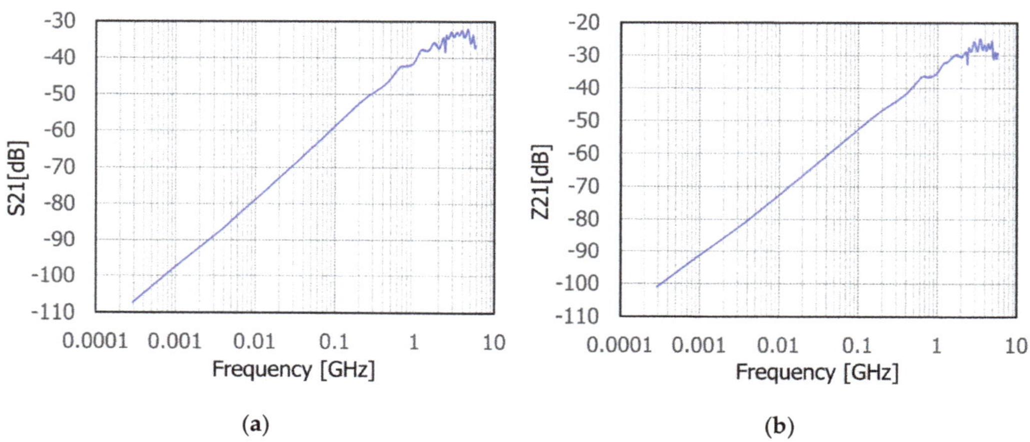

Figure 3. Coupling efficiency of designed magnetic loop probe from a microstrip line. (**a**) S21. (**b**) Z21.

Figure 4 shows the material fixture, which contains the microstrip line PCB, and the automatic probe positioning system named EME-28G [12]. The material fixture consists primarily of a test PCB holder and a cover. To ensure proper electrical contact, gaskets are placed at the interface where the two meet, and there are three push buttons each on the top and bottom. The holder houses the test PCB, i.e., the microstrip line PCB, with both ends connected to two SMA connectors on the holder. These connectors allow the material fixture to be linked to external measuring equipment via RF cables. During shielding measurements, one SMA connector is connected to one port of a vector network analyzer (VNA), while the other SMA connector is terminated with a 50 Ω SMA-type load. The automatic probe positioning system, called EMC-28G, uses three step motors to allow the probe to move along the three axes (XYZ). At the top center of the machine, a high-definition (HD) camera is mounted for observing a scanned PCB, and a probe holder is provided for securing the probe during measurements. The system is electrically and mechanically protected by a shield box.

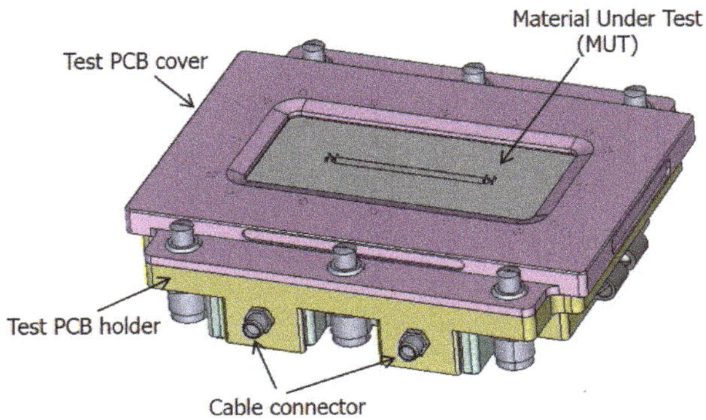

(a)

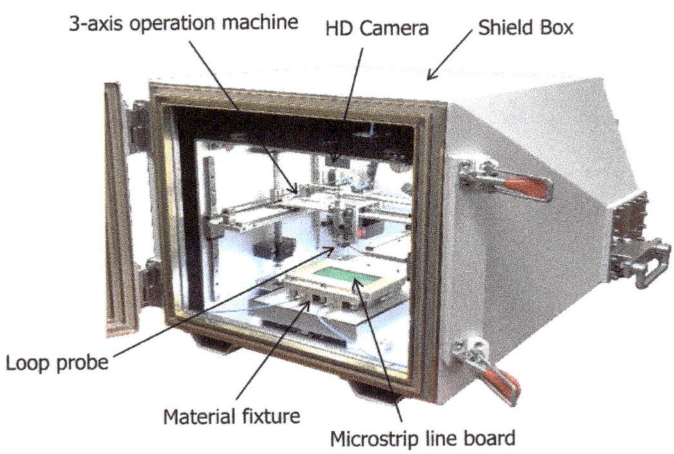

(b)

Figure 4. Fabricated apparatus for NSE measurement [11]. (**a**) Material fixture. (**b**) Automatic probe positioning system (EME-28G).

3. Measurement of Near-Field Magnetic Shielding Effectiveness (NSE)

In this section, we will explain how NSE was measured in detail. Figure 5 shows the reference measurement without a magnetic sheet and the load measurement with a sheet, using the EME-28G, a material fixture, and a loop probe. As mentioned earlier, the two SMA connectors of the material fixture are internally connected to both ends of the microstrip line. Externally, one SMA connector is connected to Port 1 of the VNA, while the other is connected to a 50 Ω termination. The rectangular loop probe, EMCIS-NH-K3-1, is positioned 3 mm above the microstrip line PCB. Using the control software that operates the EME-28G, the probe can be automatically positioned at a height of 3 mm above the center of the microstrip line. To improve the sensitivity of the received near magnetic field, a PA 306 amplifier (AMP) from Langer EMV [13], providing a 30 dB gain from 100 kHz to 6 GHz, is connected between the loop probe and the RF cable leading to Port 2 of the VNA. For the load measurement, the magnetic sheet is placed above the microstrip line in the material fixture, and the cover is applied. The two-port transmission parameter, S21, is then measured at the same probe position as in the reference measurement.

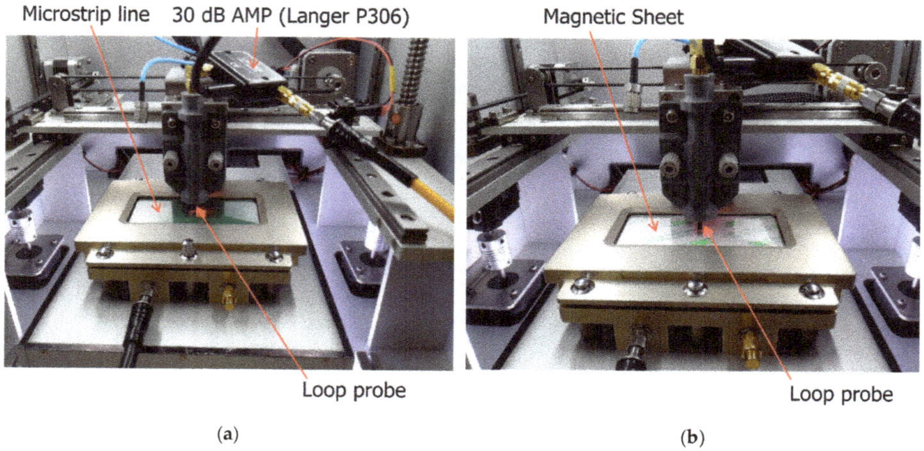

Figure 5. NSE measurement. (**a**) Reference measurement without a sheet. (**b**) Load measurement with a sheet.

Figure 6 presents the three types of commercial magnetic sheets as materials under test (MUTs). The first and second sheets, referred to as model L and model Q, respectively, are produced by the same manufacturer and have identical dimensions and thickness. Each sheet measures 100 mm × 50 mm with a thickness of 0.4 mm and is made of metal powder. Model L and model Q have a real relative permeability of 180 and 250, respectively, at 1 MHz [14,15]. According to their data sheets, both models are effective as noise suppression sheets within the 20 MHz to 3 GHz range. Despite having the same thickness, their differing permeability allow for the observation of how real permeability impacts NSE. Figure 7 shows the merged permeability data of models L and Q, which are the measured data up to 120 MHz and the extrapolated data from 120 MHz to 10 GHz. The MUTs with toroidal core shape and a thickness below 3 mm are coiled with a wire. Each relative permeability of the models L and Q was calculated by measuring the impedances of the MUTs based on the relationship between inductance and permeability [16]. The measured permeability is a little smaller than that of the data sheets [14,15]. The measured permeability was then imported into CST Microwave Studio [17], a numerical simulation software, to obtain extrapolated data up to 10 GHz. This software allows us to fit the frequency-dependent material properties with arbitrary order for the permeability by magnetic disperse fit function.

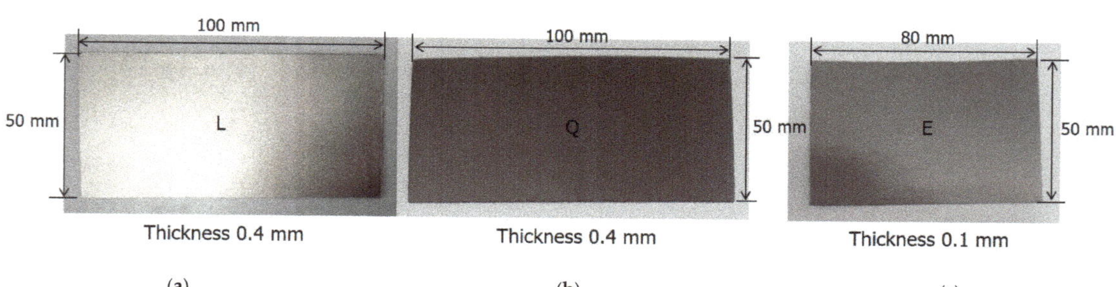

Figure 6. Three types of commercial magnetic sheets used for the measurements. (**a**) Model L. (**b**) Model Q. (**c**) Model E.

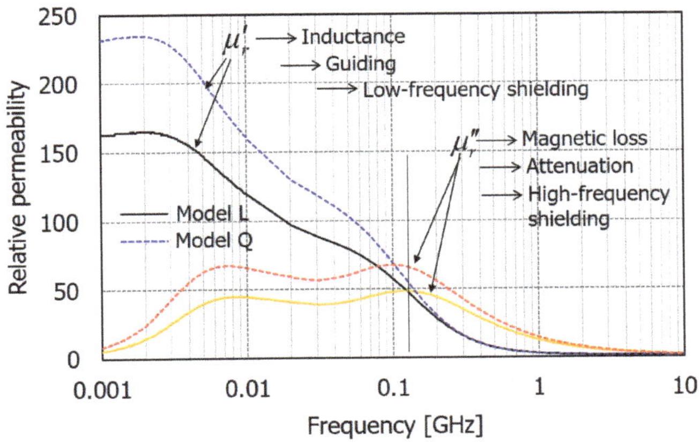

Figure 7. Relative permeability of model L and model Q.

The third sheet, model E, measures 80 mm × 50 mm with a thickness of 0.1 mm and has a relative permeability of approximately 230 at 3 MHz. According to the data sheet, this sheet is described as an effective noise suppression sheet within the 1 MHz to 3 GHz range. The size difference in model E compared to model L and model Q is due to the different manufacturers, resulting in varying sample sizes. Additionally, while model Q and model E have similar permeability, their differing thicknesses provide an opportunity to examine the effect of thickness on NSE.

Figure 8 illustrates the air gap that may exist between the microstrip line and the magnetic sheet during NSE measurements. This study aims to analyze the impact of this air gap on NSE through experimental measurements. To create a consistent air gap, three paper sheets, each with a thickness of 0.11 mm and the same dimensions as the magnetic sheet, were used as spacers, as shown in Figure 9a. If one paper sheet (spacer) is placed between the microstrip line PCB and the magnetic sheet, it creates an air gap of 0.11 mm, the same thickness as the paper. Using two paper sheets results in a 0.22 mm air gap, while using three creates a 0.33 mm air gap.

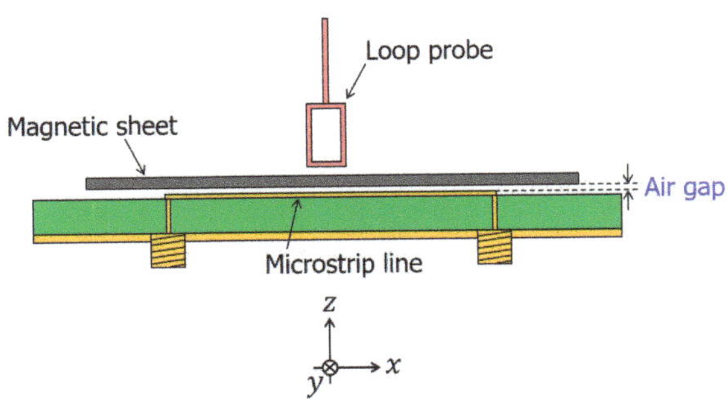

Figure 8. NSE measurement when there is an air gap between the microstrip line and the magnetic sheet.

In addition to the paper spacers, another element is needed to ensure that no additional air gaps are created beyond the thickness of the paper sheets. To achieve this, an acrylic sheet with a thickness of 1.0 mm and the same dimensions as the magnetic sheet was placed on top of the magnetic sheet, and the material fixture's cover was secured. As seen in Figure 3a, the cover of the material fixture has an opening in the center. By placing an acrylic sheet thick enough to prevent bending on top of the magnetic sheet and securing the cover, the entire magnetic sheet, not just its edges, is pressed down by the acrylic sheet, as shown in Figure 9b. This setup ensures that no extra air gaps are formed, and only the intended air gap created by the thickness of the paper sheets remains between the microstrip line PCB and the magnetic sheet.

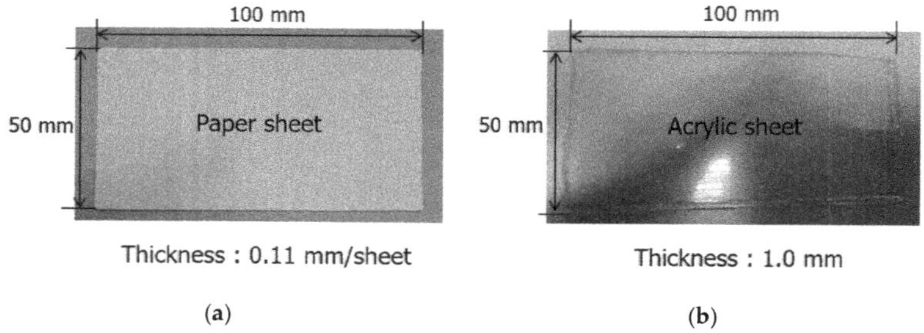

Figure 9. Photo of the paper sheet and acrylic sheet used to create a constant air gap. (**a**) Paper spacer. (**b**) Acrylic sheet.

4. Results and Discussion

Before measuring the NSE of the three magnetic sheets, the impact of the acrylic sheet and paper spacers (sheets) used to create a consistent air gap on the S21 parameter was verified. Figure 10 compares the S21 measurements of three cases including the 30 dB AMP gain. The first case is the reference measurement, where only the microstrip line is present; the second case includes an acrylic sheet placed on top of the microstrip line printed circuit board (PCB); and the third case involves three paper spacers placed between the microstrip line PCB and the acrylic sheet. The results show that when only the acrylic sheet is present, S21 is slightly higher than when only the microstrip line is present. Although a difference increases slightly with frequency, the difference is small, and the maximum difference is less than 1 dB, as shown in Figure 10. When the three paper spacers are added between the microstrip line PCB and the acrylic sheet, the results are almost identical to those of the acrylic sheet alone. Thus, it can be concluded that the paper spacers, which create the air gap, behave similarly to air and have no significant effect, and the acrylic sheet also has minimal impact on the S21 measurements.

The NSE of three magnetic sheets in Figure 6, which are used as noise suppression sheets in the GHz frequency range, was measured. First, we compared the measured S_{21}^{load} of model L at various air gaps with S_{21}^{ref} in Figure 11a. It can be seen that S_{21}^{load} increases with the number of paper spacers at frequencies above 1 GHz. Figure 11b shows the NSE, which is the difference between S_{21}^{ref} and S_{21}^{load} as presented in Figure 11a. The sheet of model L exhibits approximately 20 dB of shielding effectiveness at low frequencies, regardless of the air gap. However, as the frequency increases up to 700 MHz or 1 GHz, the shielding effectiveness decreases. Interestingly, when the frequency exceeds 1 GHz, the shielding performance of the magnetic sheet increases again at higher frequencies. This behavior is similarly observed in the NSE characteristics of models Q and E.

The frequency-dependent NSE characteristics of the magnetic sheets can be predicted by examining their permeability properties. The relative permeability of model L, shown in Figure 7, is high at low frequencies and decreases as the frequency increases. Therefore, it can be concluded that the shielding effectiveness in the low-frequency range is primarily determined by the real permeability of the magnetic sheet. Specifically, the near magnetic field generated by the microstrip line is guided into the magnetic sheet, which has a high real permeability and then makes a low reluctance path, thereby reducing the magnetic field that couples to the loop probe. As the frequency increases and the real permeability decreases, this field-guiding effect diminishes, leading to a reduction in NSE.

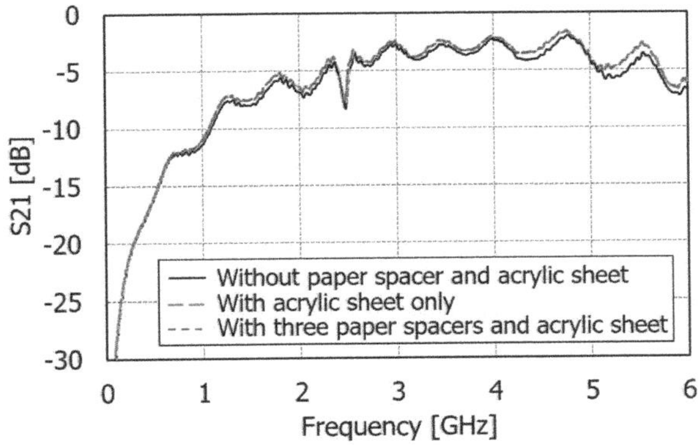

Figure 10. Effect of acrylic sheet and paper spacers on S21 measurement.

However, at higher frequencies in the GHz range, the imaginary permeability has a greater influence on NSE than the real permeability [18]. The imaginary permeability increases loss within the magnetic sheet, causing the near magnetic field generated by the microstrip line to be attenuated as it converts into losses within the sheet. Consequently, the magnetic field coupling to the loop probe is further reduced. In addition to the magnetic loss, another factor that affects the high-frequency NSE characteristics is eddy current loss. The eddy current loss occurs in magnetic sheets made of metal powder but is not present in those made of ferrite, which is an insulator. Upon examining the metal powder in the magnetic sheet, it is found to be composed of metal flakes. At high frequencies, eddy currents are induced in these flakes, increasing the loss [19,20]. This explains the frequency-dependent NSE characteristics of magnetic sheets made of metal powder.

Returning to an examination of the air gap's impact on the near-field shielding effectiveness (NSE) of model L. As shown in Figure 11a,b, in the high-frequency range above 1 GHz, S_{21}^{load} increases and NSE decreases, respectively, as the air gap increases. When a 0.11 mm thick paper spacer is inserted, NSE decreases by as much as 2 dB to 10 dB compared to the case with no air gap. Adding one more paper spacer results in an additional reduction in NSE by approximately 1 to 3 dB per spacer. At 6 GHz, when there are three paper spacers, resulting in an air gap of 0.33 mm, NSE is 20 dB lower compared to the case with no air gap. The impact of the air gap on reducing NSE becomes more pronounced as the frequency increases. This phenomenon can be explained as follows. As previously discussed, the magnetic loss and eddy current loss in the noise suppression sheet made of metal powder increase in the high-frequency GHz range, which contribute to improving NSE. However, when an air gap is introduced, the effective magnetic loss for all regions of the air gap and the magnetic sheet is reduced, leading to a decrease in NSE [21]. In conclusion, the presence of an air gap causes the NSE of the magnetic sheet to be underestimated in the GHz frequency range.

Interestingly, as shown in Figure 11b, a different trend is observed in the frequency range below approximately 700 MHz. To further investigate the NSE behavior at lower frequencies, the x-axis for frequency is presented on a logarithmic scale in Figure 11c. In the frequency range from 100 to 700 MHz, it is observed that NSE actually increases by about 1–2 dB due to the presence of the air gap. This phenomenon can be explained using the equivalent magnetic circuit composed of reluctances, as shown in Figure 12 [22,23]. Figure 12a illustrates the cross-sectional dimension of the air gap between the microstrip line and the magnetic sheet. Using the magnetic flux path shown in Figure 12b, the equivalent magnetic circuit, represented by reluctances, is also depicted in Figure 12c. Equations (1)–(4) express the individual reluctances in terms of geometrical dimensions [22,23].

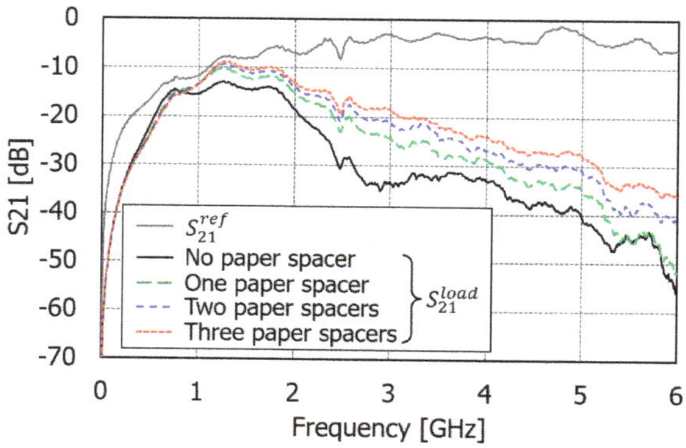

(**a**)

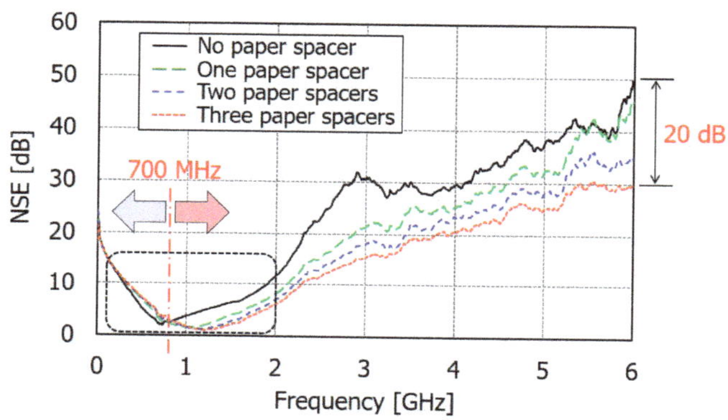

(**b**)

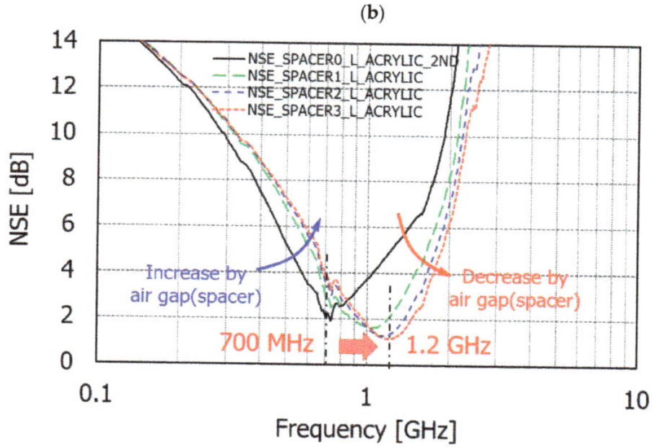

(**c**)

Figure 11. Measured S21s and NSE of model L with varying air gap. (**a**) S21s. (**b**) NSE. (**c**) NSE in log scale of *x*-axis.

$$\mathcal{R}_{a0}\ (=\mathcal{R}_{b0}) = \frac{1}{\mu_0}\left(\frac{w_s}{0.3w_s + 0.9w_m}\right), \tag{1}$$

$$\mathcal{R}_m = \mathcal{R}_m{}' + j\,\mathcal{R}_m{}'' = \left(\frac{1}{\mu_0\mu_r}\right)\left(\frac{w_s}{t}\right), \tag{2}$$

$$\mathcal{R}_{\mathrm{d0}} = \frac{1}{\mu_0}\frac{w_s}{d}, \tag{3}$$

$$\mathcal{R}_{\mathrm{ag}} = \frac{1}{\mu_0}\frac{d}{w_m}, \tag{4}$$

where w_s is the width of microstrip line and w_m is the distance from the edge of the signal line to the place where the magnetic field intensity becomes 1/e [22]. The size of the air gap can influence w_m by affecting the magnetic field distribution around the signal line (microstrip line) where current flows. However, since it affects the magnetic field at the edges rather than the top of the signal line, its impact is expected to be limited. In addition, the coefficients of w_s and w_m in (1) were obtained by fitting to the FEM simulation results [23].

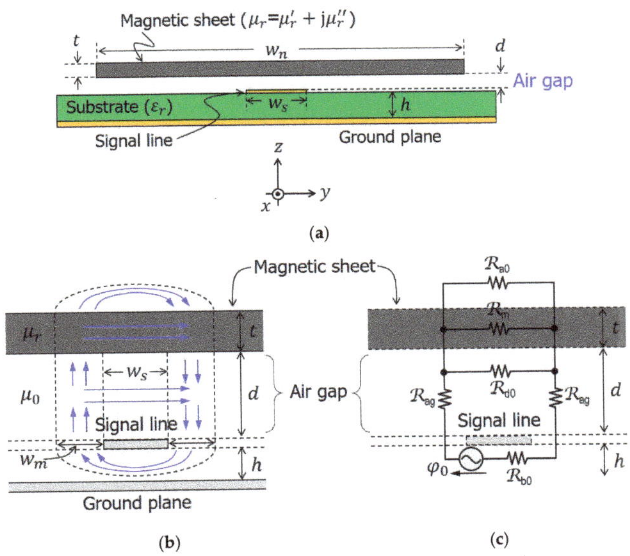

Figure 12. Cross-sectional geometry of air gap between microstrip line and magnetic sheet. (**a**) Dimensions of air gap between microstrip line and magnetic sheet. (**b**) Magnetic flux path. (**c**) Equivalent magnetic circuit represented by reluctances.

When the thickness of the air gap, denoted as d, increases, the reluctance $\mathcal{R}_{\mathrm{ag}}$ also increases, impeding the flow of magnetic flux. In addition, the reduced reluctance $\mathcal{R}_{\mathrm{d0}}$ causes part of the magnetic flux to flow not only through the magnetic sheet ($\mathcal{R}_{\mathrm{m}}$) but also through $\mathcal{R}_{\mathrm{d0}}$. This redistribution of the magnetic flux due to the air gap leads to an enhancement in shielding effectiveness. As a result, as shown in Figure 11c, the frequency at which the lowest NSE occurs increases from 700 MHz to 1.2 GHz as the air gap expands to 0.33 mm. When the frequency decreases below 100 MHz, the wavelength of the magnetic field becomes large enough that the air gap's effect can be neglected. As a result, as shown in Figure 11, NSE remains nearly identical regardless of the air gap size in this low frequency range.

Figure 13 presents the measured S_{21}^{load} and NSE of model Q as the air gap varies. Model Q is a similar product from the same manufacture as model L, and the thickness is the same. Thus, model Q exhibits S_{21}^{load} and NSE characteristics similar to those of model L, as shown in Figure 11. However, as shown in the permeability in Figure 7, model Q has a higher permeability than model L, and it is clear that the higher permeability leads to a slightly increased NSE when comparing Figure 13b with Figure 11b. At frequencies above approximately 1 GHz, NSE decreases as the air gap increases, but the rate of decrease diminishes as the air gap grows, eventually causing NSE to converge. This trend is also observed in the frequency range below 1 GHz where NSE increases due to the air gap, as shown in Figure 13c. For model Q, as the air gap increases to 0.33 mm, the frequency corresponding to the lowest NSE rises from about 680 MHz to 930 MHz. When compared to model L, the frequency with the lowest NSE has decreased slightly.

Figure 14 shows the measured S_{21}^{load} and NSE of model E as the air gap varies. Model E is from a different manufacturer than models L and Q, and it is also thinner, with a thickness of 0.1 mm. Although its permeability is greater than that of model L, it is similar to model Q in terms of permeability. However, the NSE of model E is lower than that of model L as well as that of model Q. This suggests that while higher permeability is essential for achieving better NSE, thickness is also a critical parameter. Generally, as the thickness of the magnetic sheet increases and its permeability becomes higher, more magnetic flux is concentrated in the sheet, resulting in improved NSE. The effect of the air gap on model E is similar to that observed for models L and Q. However, a distinct difference is noted in the frequency range above 1 GHz. Unlike models L and Q, where NSE continues to increase with frequency, model E experiences a drop in NSE around 2.3 GHz. For model E, as the air gap increases to 0.33 mm, the frequency corresponding to the lowest NSE shifts from approximately 1 GHz to 1.52 GHz.

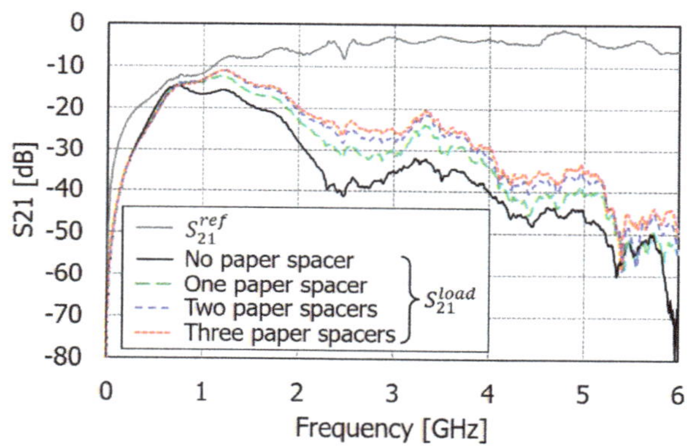

(**a**)

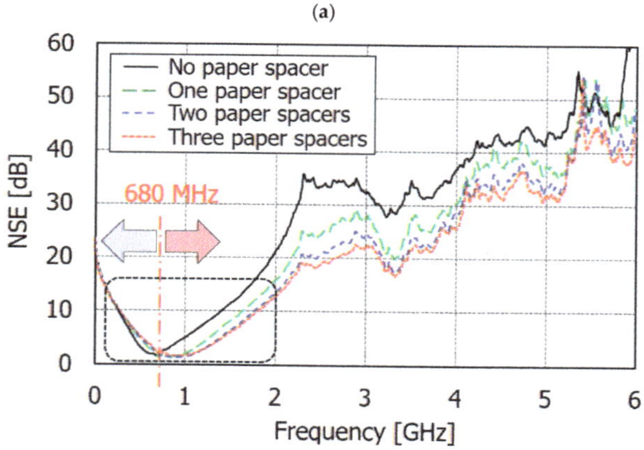

(**b**)

Figure 13. *Cont.*

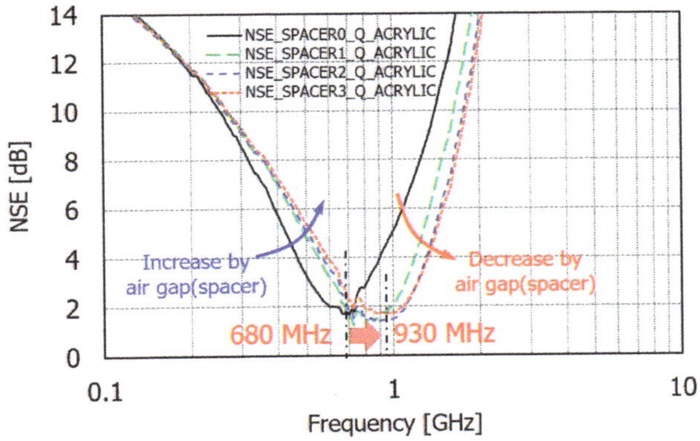

(c)

Figure 13. Measured S21s and NSE of model Q with varying air gap. (**a**) S21s. (**b**) NSE. (**c**) NSE in log scale of *x*-axis.

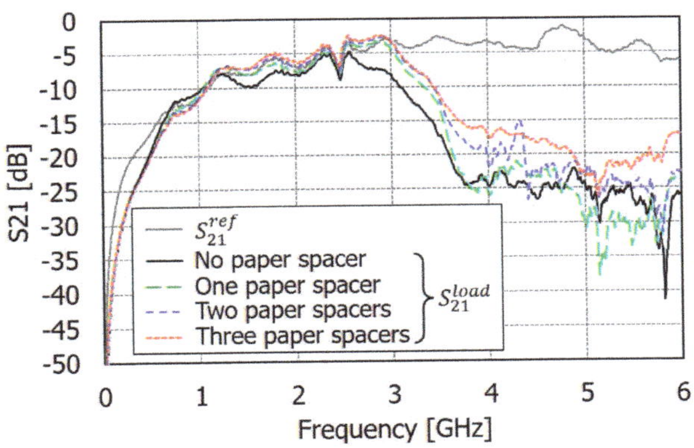

(**a**)

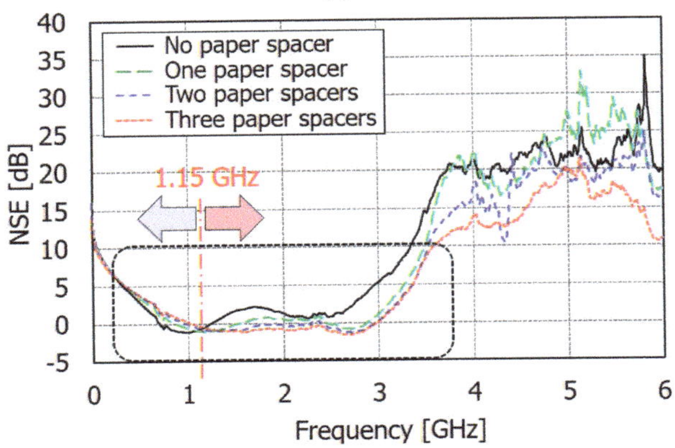

(**b**)

Figure 14. *Cont.*

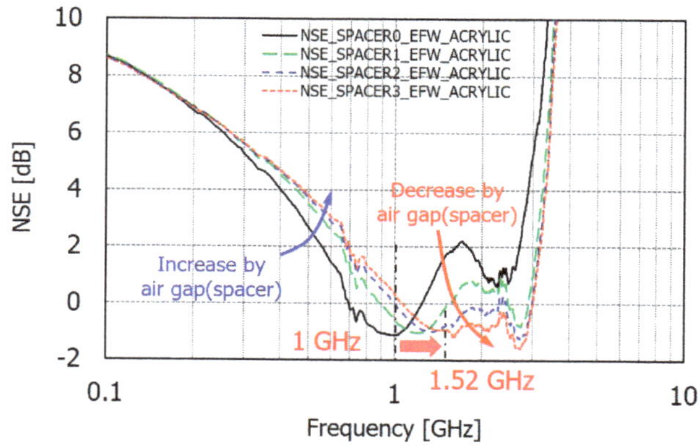

(c)

Figure 14. Measured S21s and NSE of model E with varying air gap. (**a**) S21s. (**b**) NSE. (**c**) NSE in log scale of *x*-axis.

5. Conclusions

This study successfully demonstrated the significant impact of air gaps on the near-field magnetic shielding effectiveness (NSE) of magnetic sheets made of metallic powder. Using a custom-designed measurement setup, including a microstrip line, a material fixture, a rectangular loop probe, and an automatic probe positioning system, the shielding properties were evaluated under varying air gap conditions. The results revealed that the air gap had a frequency-dependent effect on NSE: increasing the air gap enhanced shielding performance in the low frequency range but reduced it in the high frequency range. In other words, around 1 GHz, the air gap increases the near-field shielding effectiveness (NSE) of the magnetic sheets in the low frequency range, while it decreases NSE in the high frequency range. To explain this, we discussed that the influence of the real permeability of the magnetic sheets is dominant in the low frequency range, using a magnetic circuit composed of reluctances to illustrate the effect of the air gap. In contrast, in the high frequency range, the explanation was based on the losses associated with the imaginary permeability of the magnetic sheets. This research highlights the importance of controlling the air gap in practical applications to optimize the shielding performance of magnetic sheets, particularly in environments where both low and high frequencies are present. To accurately evaluate the NSE of magnetic sheets using a microstrip line and loop probe, it is crucial to properly manage the air gap between the microstrip line and the magnetic sheet during measurements. To prevent the formation of air gaps, it is important to apply an appropriate weight based on the thickness and rigidity or flexibility of the magnetic sheet under evaluation. This ensures proper contact and reliable results. The findings of this study not only provide valuable insights for evaluating the NSE of magnetic sheets but also offer practical guidance for the application of magnetic sheets in real-world electronic systems and devices. This includes how to effectively attach magnetic sheets to noise sources to reduce noise coupling and electromagnetic interference (EMI). By following the methods outlined in this research, engineers can better optimize the performance of magnetic sheets in minimizing EMI in various electronic applications.

Author Contributions: Conceptualization, H.H.P., E.S., J.K. and C.K.; methodology, H.H.P. and E.S.; validation, H.H.P. and E.S.; formal analysis, H.H.P.; investigation, H.H.P. and E.S.; resources, H.H.P., E.S. and J.K.; writing—original draft preparation, H.H.P., E.S. and J.K.; writing—review and editing, H.H.P., E.S., J.K. and C.K.; supervision, E.S., J.K. and C.K.; project administration, H.H.P. and C.K.; funding acquisition, C.K. All authors have read and agreed to the published version of the manuscript.

Funding: This research received no external funding.

Data Availability Statement: Data are contained within the article.

Acknowledgments: This work was supported by Institute of Information & communications Technology Planning & Evaluation (IITP) grant funded by the Korea government (MSIT) (No. 2020-0-00839, Development of Advanced Power and Signal EMC Technologies for Hyper-connected E-Vehicle).

Conflicts of Interest: Author C. Kim was employed by the company EMCIS, Co., Ltd. The remaining authors declare that the research was conducted in the absence of any commercial or financial relationships that could be construed as a potential conflict of interest.

References

1. Yoshida, S.; Ono, H.; Ando, S.; Tsuda, F.; Ito, T.; Shimada, Y.; Yamaguchi, M.; Arai, K.; Ohnuma, S.; Masumoto, T. High-frequency noise suppression in downsized circuits using magnetic granular films. *IEEE Trans. Magn.* **2001**, *37*, 2401–2403. [CrossRef]
2. Watanabe, K.; Jike, K.; Tanaka, S.; Miura, N.; Nagata, M.; Takahashi, A.; Miyazawa, Y.; Yamaguchi, M. Magnetic composite sheets in IC chip packaging for suppression of undesired noise emission to wireless communication channels. In Proceedings of the 2019 12th International Workshop on the Electromagnetic Compatibility of Integrated Circuits (EMC Compo), Hangzhou, China, 21–23 October 2019; pp. 219–221.
3. Deutschmann, B.; Khan, J.S.; Winkler, G.; Victoria, J. Reduction of electromagnetic emission from ICs using soft flexible ferrite sheets. In Proceedings of the 2021 13th International Workshop on the Electromagnetic Compatibility of Integrated Circuits (EMC Compo), Bruges, Belgium, 9–11 March 2022; pp. 19–24.
4. Jie, H.; Zhao, Z.; Zeng, Y.; Chang, Y.; Fan, F.; Wang, C.; See, K.Y. A review of intentional electromagnetic interference in power electronics: Conducted and radiated susceptibility. *IET Power Electron.* **2024**, *17*, 1487–1506. [CrossRef]
5. Park, H.H. Reduction of electromagnetic noise coupling to antennas in metal-framed smartphones using ferrite sheets and multi-via EBG structures. *IEEE Trans. Electromagn. Compat.* **2018**, *60*, 394–401. [CrossRef]
6. *IEC 62333-2*; Noise Suppression Sheet for Digital Devices and Equipment Part 2: Measuring Methods. 1st ed. 2006. Available online: https://webstore.iec.ch/en/publication/6868 (accessed on 15 August 2024).
7. Park, H.H.; Kwon, J.H.; Ahn, S. A simple equivalent circuit model for shielding analysis of magnetic sheets based on microstrip line measurement. *IEEE Trans. Magn.* **2017**, *53*, 9401504. [CrossRef]
8. Chen, C.-H.; Tseng, Y.-C.; Lin, I.-C.; Fu, C.-C.; Liao, K.-H.; Wu, T.-L. Transmission-line based modeling for conformal shielding in advance system-in-package (SiP). In Proceedings of the Asia-Pacific Symposium on Electromagnetic Compatibility, Taipei, Taiwan, 26–29 May 2015; pp. 521–523.
9. Chang, L.; Guo, G.; Zeng, M.; Li, E. A shielding effectiveness test system based on microstrip line. In Proceedings of the 2015 IEEE International Symposium on Antennas and Propagation & USNC/URSI National Radio Science Meeting, Vancouver, BC, Canada, 19–24 July 2015; pp. 1184–1185.
10. Park, H.H.; Kim, C. Wideband magnetic field shielding measurement method of EMI sheets using microstrip lines. *J. Korean Inst. Electromagn. Eng. Sci.* **2023**, *34*, 554–560. [CrossRef]
11. Chuang, H.-H.; Li, G.-H.; Song, E.; Park, H.-H.; Jang, H.-T.; Park, H.-B.; Zhang, Y.-J.; Pommerenke, D.; Wu, T.-L.; Fan, J. A magnetic-field resonant probe with enhanced sensitivity for RF interference applications. *IEEE Trans. Electromagn. Compat.* **2013**, *55*, 991–998. [CrossRef]
12. EMCIS Co., Ltd. Available online: https://emcis.co.kr/wpE/ (accessed on 15 August 2024).
13. PA 306 SMA; Langer EMV-Technik GmbH: Bannevitz, Germany. Available online: https://www.langer-emv.de/en/category/preamplifier/37 (accessed on 15 August 2024).
14. *EMI Noise Suppression Absorber, NoiseSorb NS1000 Series NS1000C/NS1000F/NS1000H*; Laird Technologies, Inc.: Chesterfield, MO, USA, 2020.
15. *EMI Noise Suppression Absorber, NoiseSorb NS1000 Series NS1000Q*; Laird Technologies, Inc.: Chesterfield, MO, USA, 2022.
16. Keysight Technologies. *16454A Magnetic Material Test Fixture, Operation and Service Manual*, 6th ed.; Keysight Technologies: Santa Rosa, CA, USA, 2017.
17. Microwave Studio (MWS). CST. 2020. Available online: www.cst.de (accessed on 15 August 2024).
18. Barba, A.; Clausell, C.; Jarque, J.C.; Nuño, L. Magnetic complex permeability (imaginary part) dependence on the microstructure of a Cu-doped Ni–Zn-polycrystalline sintered ferrite. *Ceram. Int.* **2020**, *46*, 14558–14566. [CrossRef]
19. Shi, K.; Zhang, D.; Ren, Z.; Chen, D.; Zhang, Y.; Koh, C.-S. Effect of particles size on eddy-current loss of soft magnetic composites using combined DEM and FEM. *IEEE Trans. Magn.* **2024**, *60*, 2000308. [CrossRef]
20. Pošković, E.; Ferraris, L.; Franchini, F.; Grande, M.A. The effect of particle size on the core losses of soft magnetic composites. *AIP Adv.* **2019**, *9*, 035224. [CrossRef]
21. Jez, R. Influence of the distributed air gap on the parameters of an industrial inductor. *IEEE Trans. Magn.* **2017**, *53*, 8401605. [CrossRef]

22. Mikami, T.; Muroga, S.; Tanaka, M.; Endo, Y.; Hashi, S.; Ishiyama, K. Estimation of noise suppression in MSL with Co-Zr-Nb film considering impedance matching. *IEEE Trans. Magn.* **2022**, *58*, 6100205. [CrossRef]
23. Mikami, T.; Muroga, S.; Tanaka, M.; Endo, Y.; Naoe, M.; Hashi, S.; Ishiyama, K. Estimation of material characteristics of film-type noise suppressor using equivalent circuit modeling and genetic algorithm. *IEEE Trans. Magn.* **2023**, *59*, 9201304. [CrossRef]

Disclaimer/Publisher's Note: The statements, opinions and data contained in all publications are solely those of the individual author(s) and contributor(s) and not of MDPI and/or the editor(s). MDPI and/or the editor(s) disclaim responsibility for any injury to people or property resulting from any ideas, methods, instructions or products referred to in the content.

electronics

Article

Analysis of Shielding Performance in Double-Layered Enclosures with Integrated Absorbers

Jong Hwa Kwon [1], Chang-Hee Hyoung [1] and Hyun Ho Park [2,*]

1 Radio Research Division, Electronics and Telecommunications Research Institute, Daejeon 34129, Republic of Korea; hjkwon@etri.re.kr (J.H.K.); hyoung@etri.re.kr (C.-H.H.)
2 School of Electrical and Electronic Engineering, The University of Suwon, Hwaseong 18323, Republic of Korea
* Correspondence: hhpark@suwon.ac.kr

Abstract: Generally, various technologies, including waveguide below cutoff (WBC), gasket sealing, and bonding, are employed in metallic enclosures to achieve the high electromagnetic shielding performance required for EMP protection and EMC countermeasures in shielding structures or facilities. While the shielding structure or facility is properly constructed and maintained according to design specifications, its electromagnetic shielding performance can remain at the required level, effectively protecting internal electrical and electronic equipment from external electromagnetic interference. However, unintended apertures often occur during the construction or maintenance of shielding facilities, compromising their shielding performance. Therefore, it is crucial to develop technologies that prevent shielding effectiveness degradation caused by both intentional and unintentional apertures. This paper proposes a structure incorporating a composite absorber (made of dielectric and magnetic absorber) within a double metal panel of enclosure featuring an aperture, aimed at maintaining and improving the facility's shielding performance. The effectiveness of the proposed structure was validated through numerical simulation.

Keywords: shielding effectiveness; metallic enclosure; absorber; aperture; electromagnetic compatibility

1. Introduction

With the rapid advancement of electrical and electronic technologies, along with information and communication technologies (ICT), many critical infrastructures in various countries are increasingly being remotely controlled or operated automatically. In this context, the social impact of high-power electromagnetic threats and attacks, including intentional electromagnetic interference (IEMI) [1], becomes significantly pronounced when major facilities are targeted. As a result, there is a growing interest in and need for protective technologies to safeguard essential equipment and systems against high-power electromagnetic threats.

Metallic enclosures are commonly used to provide electromagnetic shielding and mechanical protection for internal equipment and systems against both intentional and unintentional electromagnetic interference, such as unwanted noise or electromagnetic pulses. However, these enclosures often feature intentional apertures for functional purposes, such as cable routing, ventilation, or access points, as well as unintentional seams that may arise during manufacturing or assembly. These apertures can compromise the enclosure's shielding effectiveness, leading to a potential reduction in its protective capabilities [2–12].

In practical applications, various types of holes or joints in the enclosure, such as those for air ventilation, access doors, and cabling, are inevitable. However, these apertures degrade the shielding performance needed to protect critical equipment and systems from external electromagnetic fields. To ensure complete protection against electromagnetic penetration, the openings and slots in the metal enclosure must be properly managed and addressed [13–15].

To date, the mechanism of electromagnetic coupling through apertures in metallic enclosures has been investigated using both numerical and experimental analyses to minimize the impact of these openings on shielding effectiveness. Research has focused on resistive sheets and electromagnetic bandgap structures installed around the slots to enhance the enclosure's shielding performance without compromising the ventilation capabilities of the apertures [16]. Additionally, the use of wire arrangements across the slots has been proposed as a cost-effective method to attenuate and disrupt slot resonance [17]. These methods can only be applied if the locations of the openings or slots are already known. However, identifying unwanted apertures in shielding structures can be challenging, particularly in large shielding chambers or equipment. When an electromagnetic field penetrates a large metal enclosure, it generates multiple resonance modes due to standing waves that are continuously reflected from the enclosure's metallic walls. These standing waves and resonances increase the electric field strength at the resonant frequency, degrading shielding performance and ultimately negatively impacting the internal devices within the shielding facility. In addition, at the resonant frequency, the uniformity of the electric field inside the enclosure diminishes, and the shielding effectiveness (SE) level varies significantly based on the position within the enclosure. Therefore, it is essential to control and enhance the shielding performance of large metallic enclosures with unknown openings. By reducing the Q-factor of the enclosure, the field penetrating the enclosure becomes less influenced by its resonance, preventing an increase in the internal electric field strength. This phenomenon can positively impact the shielding properties of the enclosure. If there are no apertures in the shielding structure or facility, it can achieve a high level of shielding effectiveness exceeding 100 dB. However, even with the presence of unintentional apertures, maintaining a certain level of shielding performance can enhance the reliability of the shielding structure or facility and significantly reduce construction or maintenance costs.

In this paper, a structure is proposed that incorporates a composite absorber (either dielectric or magnetic absorber) placed within a dual-metal-panel enclosure featuring an aperture, aimed at enhancing the shielding performance of the facility. The effectiveness of this structure was verified through numerical simulation. The proposed structure has been confirmed to maintain a certain level of shielding effectiveness even in the presence of arbitrary intentional and unintentional apertures in both dual shielding panels. Therefore, it is suitable for electromagnetic protection facilities that require a minimum level of shielding performance, such as those designed to protect critical equipment from high-power electromagnetic threats.

2. SE of Enclosure with Aperture

The shielding performance of the conductive enclosures is determined by three elements, as illustrated in Figure 1. The first one is the effect of the material itself and the other two are the effects of the enclosure's structure. Shielding by reflection (R), absorption (A), and multiple reflection (B) is the effect of the material itself, while shielding by leakage through aperture and resonance is the effect of the structure [8–10,13].

$$\mathrm{SE} = \underbrace{\mathrm{A+R+B}}_{\text{Material Effects}} - \underbrace{\overbrace{\text{Leakage Effect}}^{\text{Aperture Effect}} - \overbrace{\text{Standing wave Effect}}^{\text{Enclosure Resonacne Effect}}}_{\text{Structure Effects}} \quad (1)$$

where A represents the effect due to absorption by the material, R denotes the effect due to reflection at the material's surface, and B refers to the impact of multiple reflection losses [11].

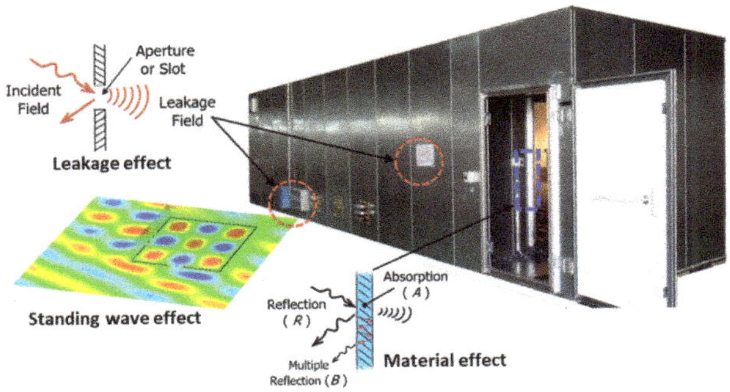

Figure 1. Electromagnetic shielding mechanism of practical metallic enclosures with apertures [8–10].

This section investigates the shielding effectiveness of enclosures with apertures through simulation. Figure 2 illustrates the structures used to analyze the shielding effectiveness of an enclosure with an aperture of size $l \times 10$ mm^2. Figure 2a shows a structure with a vertically oriented aperture placed on an infinite metallic plane, while Figure 2b depicts a vertically oriented aperture located on one side of a metallic enclosure. The incident electromagnetic wave is a plane wave with horizontal polarization, striking either the infinite plane or the enclosure perpendicularly. In Figure 2, the probe position, where the electric field strength and shielding effectiveness were calculated, is located 4.5 m away from the aperture in the horizontal direction for the infinite plate. In the case of the shielding enclosure, the observation point is the center of the enclosure, where the distance from the aperture remains the same at 4.5 m. The electric field strength at the observation positions was analyzed using a numerical analysis tool CST MWS [18], and the shielding effectiveness was subsequently calculated. The consistency of the simulation results for metallic enclosure structures obtained using CSW MWS with the measured shielding effectiveness has been validated in previous research [10].

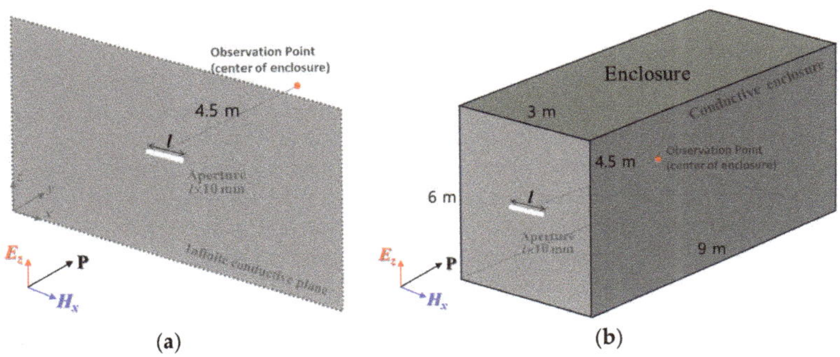

Figure 2. Structures of infinite conductive plane and enclosure with an aperture: (**a**) Aperture on infinite metallic plates; (**b**) Metallic enclosure with an aperture.

Figure 3 illustrates the trends in shielding effectiveness for an infinite plane and an enclosure with an aperture. The solid line represents the shielding effectiveness of the infinite plane, while the dotted line indicates the effectiveness of the enclosure. As shown in Table 1, the shielding effectiveness of the infinite metallic plane decreases up to the resonance frequency range of the aperture. Although the overall trend for the enclosure with the same aperture is similar to that of the infinite plane, variations arise from the enclosure's resonances, ultimately leading to reduced shielding effectiveness.

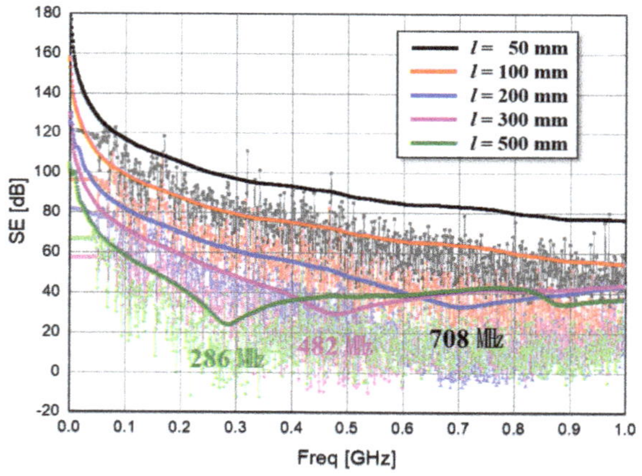

Figure 3. Analysis results of shielding effectiveness for infinite plane and enclosure with aperture.

Table 1. Resonant frequencies according to aperture length.

Length [mm]	50	100	200	300	500
Resonance Frequency [MHz]	3000	1500	750	500	300

In practical metallic structures, any form of aperture permits electromagnetic fields to enter, which can lead to resonance within the enclosure. Furthermore, the shape and size of the aperture can induce its own resonance, resulting in increased electromagnetic fields penetrating the enclosure around the resonant frequency range. To mitigate the impact of electromagnetic fields caused by the structure, it is essential to minimize the number of apertures and reduce the standing waves generated inside.

3. Proposed Structure and Analysis

This paper proposes a "dual-metal-panel shielding structure with embedded composite absorbers" that maintains a certain level of shielding performance even when apertures are present in the shielding structure. As illustrated in Figure 4, the proposed design consists of an enclosure with dual shielding panels, each featuring apertures, allowing for an examination of whether shielding effectiveness is preserved in the presence of unintended apertures. Three apertures of the dual-metal-panel shielding structure are located at different positions on the inner and outer metal panels. The size of the metallic enclosure is $0.8 \times 0.6 \times 0.5$ m^3, and the dimensions of the horizontally positioned aperture on the panel exposed to the external electromagnetic field are 100×10 mm^2, with a conductor thickness of 3 mm. The gap between the dual metal panels is denoted as d mm, and the distance between the apertures on the inner and outer panels is 125 mm, measured from the center of the apertures.

In this study, to account for the worst-case shielding effectiveness scenario, the plane wave was configured to incident perpendicularly on the surface containing apertures, with the electric field polarization oriented perpendicular to the maximum length direction of the aperture. To assess the electromagnetic shielding effectiveness of the enclosure with apertures, a plane wave was employed as the incident electric field, specifically using vertical polarization to examine the effects of the apertures. Since the apertures are oriented horizontally on the enclosure, the vertical component of the electric field results in the lowest shielding characteristics. The analysis frequency range is from 1 MHz to 3 GHz, and the electric field at the center of the enclosure was analyzed using numerical simulation tools, allowing for the calculation of shielding effectiveness. The parameters used in the simulations considering the worst-case scenario are provided in Table 2.

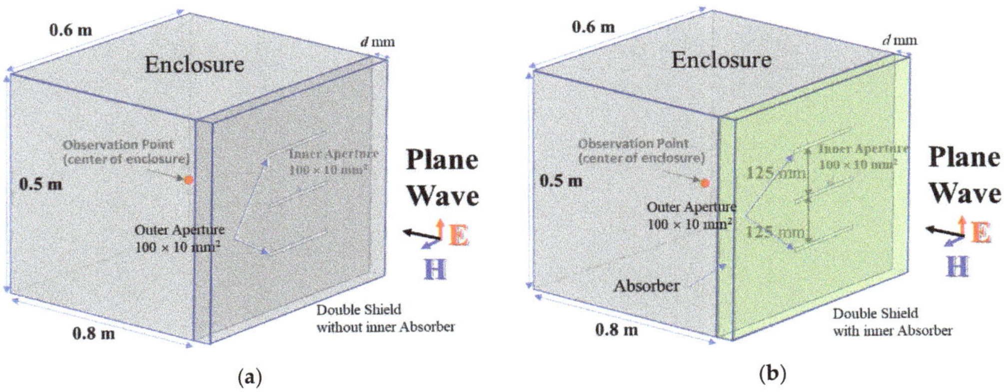

Figure 4. Enclosure structure with apertures at different positions on inner and outer metal panel: (**a**) Dual-metal-panel enclosure without absorber; (**b**) Dual-metal-panel enclosure with absorber.

Table 2. Parameters for calculating the shielding effectiveness of enclosure with absorber.

Simulation Parameter		Value
Size of Enclosure		$0.6 \times 0.5 \times 0.8$ m^3
Thickness of Enclosure's Metal Panel		3 mm
Size of Aperture		0.1×0.01 m^2
Absorber	Dielectric Absorber	Eccorsorb LS Series [18]
	Magnetic Absorber	FSA300 [19], FS170 [20] & MF190 [21]
Observation Point for Electric Field		Center of Enclosure
Frequency Range		0.001~3 GHz

In this study, shielding performance was enhanced using a composite absorber, consisting of a dielectric absorber and a magnetic absorber, to improve the shielding effect across a broader frequency range. Laird's product [19] was used as the dielectric absorber, while magnetic absorbers from Laird, GSS [20], and DMEGC [21] were employed to enhance shielding performance in the low-frequency range. The complex permittivity and permeability of the Laird absorber are provided in CST MWS, while the complex permeability of the GSS and DMEGC magnetic absorbers was measured and is presented in Figure 5. To assess the effectiveness of the absorbers in improving shielding performance, numerical analysis was conducted, evaluating different absorber types, thicknesses, and arrangement methods.

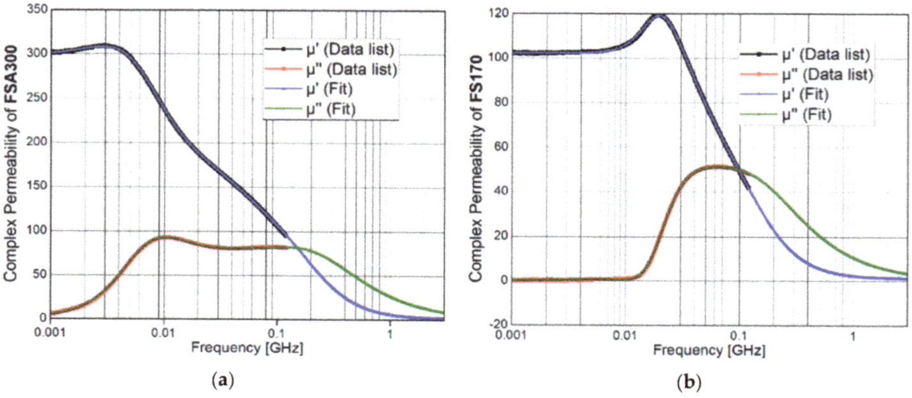

Figure 5. Complex permittivity of magnetic absorber [7]: (**a**) FSA300; (**b**) FS170.

4. Simulation and Result Analysis

In this paper, the shielding effectiveness of a dual-metal-panel enclosure with an arbitrary aperture, to which a composite absorber comprising both dielectric and magnetic materials is applied, was analyzed using CST MWS [18]. Additionally, Equation (2) was employed to assess the shielding effectiveness when the absorber is placed within a dual-metal-panel enclosure featuring an aperture, compared to the scenario where no metallic enclosure is present.

$$\mathrm{SE} = 20\log\left(\frac{E_{w/o\ Enc}}{E_{Enc\ w\ Abs}}\right) \quad (2)$$

where $E_{w/o\ Enc}$ represents the electric field intensity at the center of the enclosure when no enclosure is present, while $E_{Enc\ w\ Abs}$ denotes the electric field intensity within the enclosure when an absorber is embedded within a dual-metal-panel enclosure. Since a plane wave is employed as the incident wave, the electric field value, $E_{w/o\ Enc}$ is set to 1 V/m when there is no enclosure.

In this section, the electric field inside a dual-metal-panel shielding enclosure with an aperture of the same dimensions (100×10 mm^2) was analyzed, with the apertures positioned at different locations on the inner and outer panels of the dual-panel enclosure, as shown in Figure 4. The shielding effectiveness was calculated using Equation (2). Additionally, the shielding effectiveness was compared to the case where the aperture of the same size was located at the center of a single-metal-panel enclosure, serving as a reference.

4.1. Dielectric Absorbers

To enhance the shielding effectiveness of enclosures with apertures, dielectric absorbers, which are commonly used as countermeasures against electromagnetic interference, were employed in this study. Figure 6 analyzes the impact of the type of absorber (Laird's Eccosorb: LS20, LS22, LS24, and LS26) and thickness (specifically LS26 at $d = 10$ mm, 20 mm, and 30 mm) when the dielectric absorber is placed inside a dual shielding structure with apertures. The insertion loss of the Eccosorb absorber [19] follows the order LS20 > LS22 > LS24 > LS26, with the loss increasing as the model number increases.

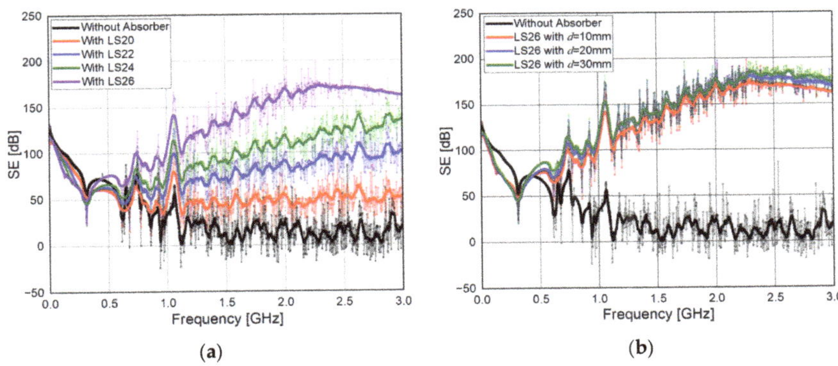

Figure 6. Shielding property of the dual-metal-panel enclosure with dielectric absorber: (**a**) Effects of absorber type with thickness of d = 10 mm; (**b**) Effects of absorber thickness.

Figure 6a illustrates the shielding effectiveness based on the type of absorber. When the apertures are positioned differently on the inner and outer metal panels, the electric field entering through the apertures of the outer shielding panel passes through the absorber to the aperture of the inner panel, resulting in a reduction effect corresponding to the insertion loss of the absorber. Consequently, a greater insertion loss of the absorber leads to a more significant reduction in the electric field. Additionally, as the frequency increases, the relative distance becomes longer, further diminishing the electric field. When LS20, LS22, LS24, and LS26 are applied in the GHz band, the approximate levels of electromagnetic attenuation are around 50 dB, 70~100 dB, 80~140 dB, and 100~180 dB, respectively.

Figure 6b illustrates the shielding effectiveness based on the thickness (d) of the absorber. The thickness of the absorber, or the spacing between the inner and outer panels, does not appear to significantly impact the shielding effectiveness. In the case of a dual-metal-panel shielding structure with apertures, the electromagnetic field incident through the aperture on the outer panel propagates into the dual-metal-panel shielding structure and enters the interior of the enclosure through the aperture on the inner panel. The electromagnetic field that enters through an aperture on the outer panel is attenuated by the absorber within the structure. Since the electromagnetic field is attenuated by the absorber located between the inner and outer apertures, variations in absorber thickness have a minimal impact on the overall shielding performance.

However, the reduction effect remains relatively insignificant below GHz. This is attributed to the fact that the absorber used is designed to operate in the GHz band, resulting in minimal loss in the frequency range below GHz.

4.2. Magnetic Absorbers

To enhance the shielding effectiveness in frequencies below the GHz range, a magnetic absorber designed to operate at low frequencies was proposed. In the dual-metal-walled shielding structure with apertures, when the positions of the apertures on the inner and outer panels differ, a magnetic absorber is placed inside the structure, and the electric field strength is measured at the center of the enclosure. The absorbers used in the analysis include magnetic types (FSA300, FS170, and MF190), each with a thickness of 20 mm, which corresponds to the spacing of the dual-metal-panel shielding structure. The permeability of the magnetic absorbers used in this study was highest for FSA300 [20], followed by FS170 [21] and MR190 [19]. As shown in Figure 7a, it was confirmed that the loss induced by the magnetic absorbers is proportional to their permeability.

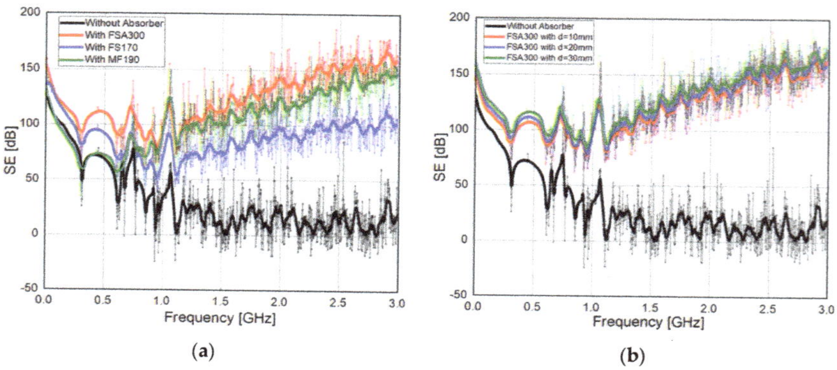

Figure 7. Shielding effectiveness of the dual-metal-panel enclosure with magnetic absorber: (**a**) Effects of absorber type with thickness of d = 20 mm; (**b**) Effects of absorber thickness.

Figure 7a illustrates the shielding effectiveness at the center of the enclosure based on the type of magnetic absorber used. The application of a magnetic absorber enhances the attenuation of the electromagnetic field in the low-frequency band. Particularly, the reduction level at the first resonant frequency (300 MHz) of the enclosure varies by absorber type: approximately 69 dB for FSA300, 58 dB for FS170, and about 35 dB for MF190. Interestingly, when magnetic absorbers are used, the SE increases significantly in the frequency range below 1 GHz, compared to dielectric absorbers as shown in Figure 6a. This is due to the higher permeability of the magnetic absorbers, as illustrated in Figure 4.

Figure 7b illustrates the variation in shielding effectiveness based on the thickness of the magnetic absorber placed inside the dual metal walled shielding structure. FSA300 was used as the absorber, with thicknesses adjusted in 10 mm increments from 10 mm to 30 mm. As shown in the figure, when the absorber is positioned in a dual-metal-panel enclosure with apertures at different locations, changes in absorber thickness, similar to those observed with dielectric absorbers, do not appear to significantly affect the electromagnetic field attenuation characteristics.

4.3. Composite Absorbers

From the previous analysis results, it was found that dielectric absorbers improve electromagnetic field reduction effects in frequency ranges above GHz, while magnetic absorbers show improvement in frequency ranges below GHz, particularly at resonant frequencies. Reflecting these analysis results, a structure incorporating a composite absorber composed of both dielectric and magnetic absorbers was proposed, and the shielding effectiveness was analyzed. The GHz band absorber used was Eccosorb LS 26 provided by Laird, while magnetic absorbers FSA300, FS170, and RF190 were employed. The spacing of the dual-metal-panel shielding structure was set at 20 mm, with absorbers applied at a thickness of 10 mm each.

Figure 8 shows the shielding effectiveness at the center of the enclosure when a plane wave with vertical polarization is incident on a surface with an aperture. The application of the composite absorber demonstrates a reduction effect of at least approximately 50 dB across the entire frequency range, including the low-frequency band. This trend is similar to the results shown in Figure 7, where only the magnetic absorber (FSA300 at 20 mm) was applied. Specifically, considering the thickness of the magnetic absorber (10 mm) for FSA300, the reduction performance in the resonant frequency range below GHz is relatively small. However, in the GHz band, the composite absorber (LS26 at 10 mm and FSA300 at 10 mm) shows slightly improved performance compared to FSA300 at 20 mm. This suggests that in the GHz band, LS26 contributes to a greater reduction of electromagnetic waves. This study indicates that both dielectric absorbers (for the high-frequency range) and magnetic absorbers (for the low-frequency range) must be applied simultaneously in the proposed structure to enhance the shielding effectiveness across a broader frequency range.

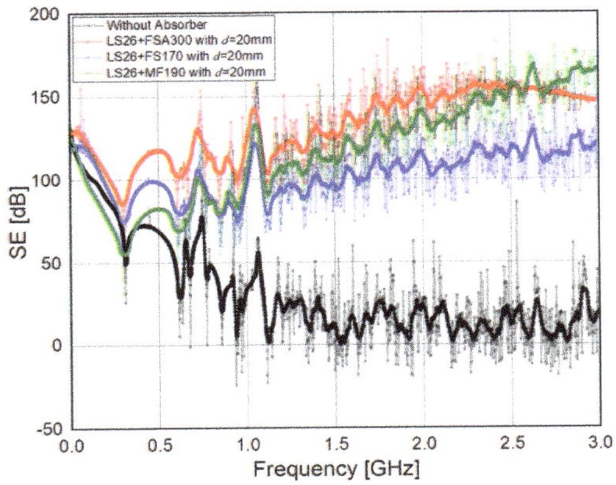

Figure 8. Analysis results of the dual-metal-panel enclosure based on type of composite absorber with total thickness of 20 mm.

4.4. SE According to Distance Between Apertures

In the dual-metal-panel structure, it is necessary to confirm that the shielding effectiveness is primarily influenced by the distance between the apertures rather than the thickness of the inner absorber. To this end, the effect of the distance between the apertures on the inner and outer panels, as shown in Figure 9, was analyzed by varying the distance between the apertures in the outer conductor of the double shielding panel structure. This analysis aimed to interpret its impact on shielding effectiveness.

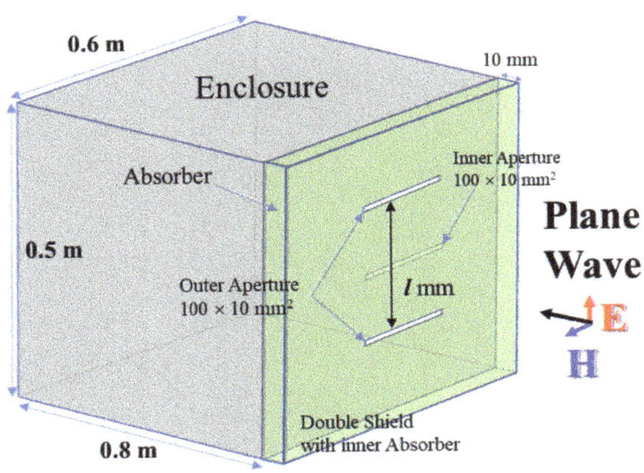

Figure 9. Enclosure with varying distances between the apertures on the outer metal panel.

The variation in the electric field inside the enclosure was analyzed based on the distance between the apertures located on the outer panel of the dual-metal-panel structure, while the aperture on the inner panel remained fixed at the center. Eccosorb LS 26, with a thickness of 10 mm, was used as the embedded absorber. The distances (l) between the apertures on the outer panel were examined for configurations of 50 mm, 100 mm, 150 mm, and 200 mm.

Figure 10 illustrates the shielding effectiveness at the center of the enclosure in relation to the varying distance between the apertures on the outer panel of the dual-metal-panel shielding enclosure. As the distance between the apertures increases, the shielding effectiveness improves significantly. This indicates that when an aperture is present at a different location on the inner and outer shielding panels, the electric field entering through the aperture of the outer shielding panel travels through the absorber to the aperture of the inner shielding panel. Consequently, as the distance between the outer and inner apertures increases, the electric field strength inside the enclosure decreases, leading to an enhanced shielding performance.

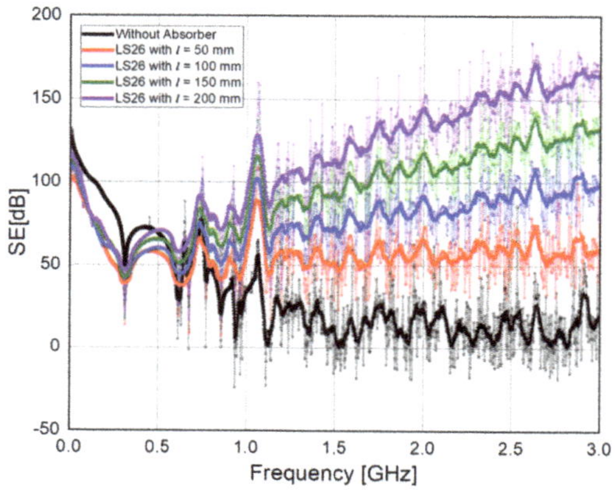

Figure 10. Analysis results of dual-metal-panel enclosure due to varying distances between the apertures on the outer metal panel.

Next, the electric field inside the enclosure was calculated with the apertures positioned at the inner and outer panels, as shown in Figure 11, while adjusting the distance between the two apertures based on the center of the enclosure surface. Eccosorb LS 26 was used as the embedded absorber in the dual-metal-panel shielding structure. The analysis focused on configurations where the distance (l) between the openings on the inner and outer panels was set to 0 mm, 50 mm, 100 mm, and 150 mm.

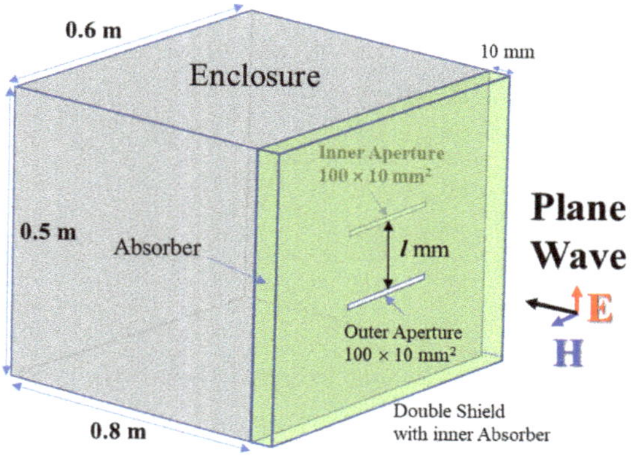

Figure 11. Enclosure with varying distances between apertures on the outer and inner metal panels.

Figure 12 shows the shielding effectiveness at the center of the enclosure in relation to the varying distances between the apertures on the inner and outer panels of the dual-metal-panel structure. Consistent with the results presented in Figure 10, the shielding effectiveness appears to increase significantly as the distance between the apertures widens.

Specifically, as the distance between the outer and the inner apertures increases, the electric field strength inside the enclosure decreases, resulting in improved shielding effectiveness.

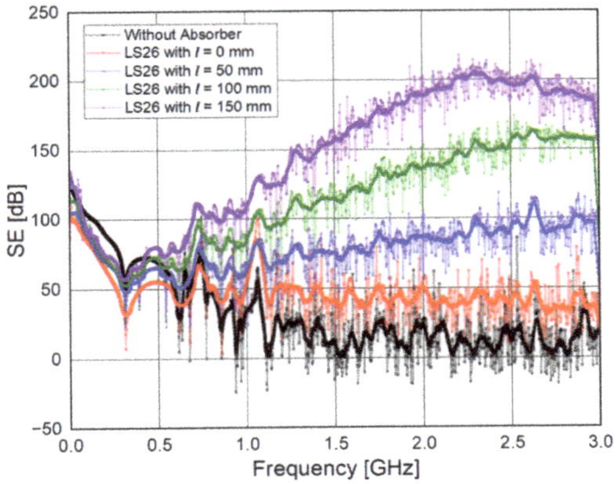

Figure 12. Enclosure with varying distances between the apertures on the outer metal panel.

However, if the apertures on the inner and outer panels are aligned, the electric field entering the aperture of the outer panel passes through the relatively thin absorber and enters the enclosure through the aperture of the inner panel. This configuration leads to resonance, resulting in diminished shielding effectiveness that is comparable to that of a single-panel enclosure structure.

In actual shielding facilities, multiple conductive panels are connected to create the structure. During construction, inadequate electrical connections between the panels can result in gaps, and over time, elongated slot-shaped gaps may develop at the junctions between the panels. It is confirmed in Figures 10 and 12 that the application of dual metal panels with embedded composite absorbers in these shielding structures can maintain a significant level of shielding effectiveness.

5. Conclusions and Discussion

Apertures are widely recognized as essential components in conductive enclosures, serving various purposes such as air ventilation, power and signal cabling, and operator access. When an electromagnetic field interacts with an enclosure that has apertures, it penetrates through these openings and interacts with the enclosure's resonance properties, generating an internal electromagnetic field. In metallic enclosures, even minimal electromagnetic transmission can induce a significant electric field at resonant frequencies, posing a potential threat to internal equipment. To mitigate this issue, it is advisable to first minimize the number of apertures and subsequently adjust the resonance properties of the enclosure using electromagnetic absorbers.

In this paper, a dual-metal-panel shielding enclosure incorporating a composite absorber composed of dielectric and magnetic materials is proposed to maintain and enhance the shielding performance of the facility. The effectiveness of this structure was verified through simulation.

While the mechanism by which the electromagnetic field, incident through the external aperture, causes resonance within the shielding structure and subsequently enters the interior of the shielding enclosure through the internal aperture remains the same, the presence of the absorber inside the dual-metal-panel shielding structure significantly reduces the strength of the electromagnetic waves. As a result, the intensity of the electromagnetic field entering the shielding enclosure is considerably diminished.

The use of dual-metal-panel shielding structures with embedded absorbers is more costly than single-metal-panel designs. However, this approach provides more reliable shielding performance. Required effectiveness can be maintained as long as unintentional apertures do not coincide. If they do, increasing the absorber's loss or thickness can still achieve the desired shielding. To further enhance effectiveness while minimizing the impact of unintended apertures, dividing the dual-panel structure into manageable sections can improve performance and facilitate maintenance. Staggering the shielding plates during fabrication may also prevent coinciding apertures, enhancing the overall performance of the dual-shielding structure.

Further research is needed to enhance the shielding effectiveness of conductive enclosures with apertures, focusing on exploring different absorber configurations and improving the performance of the absorbers. To this end, simulations will be conducted to analyze various structural configurations, including different types and placements of apertures, as well as the types and arrangements of absorbers. Additionally, the shielding effectiveness of metallic enclosures with apertures will be further evaluated through experimental measurements.

Author Contributions: Conceptualization, J.H.K. and H.H.P.; methodology, J.H.K. and H.H.P.; software, J.H.K. and H.P; validation, J.H.K., C.-H.H. and H.H.P.; formal analysis, J.H.K. and H.H.P.; investigation, J.H.K. and H.H.P.; data curation, J.H.K., C.-H.H. and H.H.P.; writing—original draft preparation, J.H.K. and H.H.P.; writing—review and editing, J.H.K. and H.H.P.; project administration, J.H.K. and C.-H.H.; funding acquisition, J.H.K. All authors have read and agreed to the published version of the manuscript.

Funding: This research was supported by the Institute of Information & Communications Technology Planning & Evaluation (IITP) grant funded by the Korean government (MSIT) (No. 2020-0-00917, Development of Key Technologies for Low-cost EMP Protective Materials, Components, Devices (MCD) & Facilities Vulnerability Assessment).

Data Availability Statement: Data are contained within the article.

Acknowledgments: This research was supported by the Institute of Information & Communications Technology Planning & Evaluation (IITP) grant funded by the Korean government (MSIT) (No. 2020-0-00917, Development of Key Technologies for Low-cost EMP Protective Materials, Components, Devices (MCD) & Facilities Vulnerability Assessment).

Conflicts of Interest: The authors declare no conflicts of interest.

References

1. Jie, H.; Zhao, Z.; Zeng, Y.; Chang, Y.; Fan, F.; Wang, C.; See, K.Y. A review of intentional electromagnetic interference in power electronics: Conducted and radiated susceptibility. *IET Power Electron.* **2024**, *17*, 1487–1506. [CrossRef]
2. Li, M.; Ma, K.-P.; Drewniak, J.L.; Hubing, T.H.; Van Doren, T.P. Numerical and experimental corroboration of an FDTD thin-slot model for slots near corners of shielding enclosures. *IEEE Trans. EMC* **1997**, *39*, 225–232.
3. Cerri, G.; de Leo, R.; Primiani, V.M.; Righetti, M. Field penetration into metallic enclosures through slots excited by ESD. *IEEE Trans. EMC* **1994**, *36*, 110–116. [CrossRef]
4. Criel, S.; Martens, L.; De Zutter, D. Theoretical and Experimental Near-Field Characterization of Perforated Shields. *IEEE Trans. EMC* **1994**, *36*, 161–168. [CrossRef]
5. Xia, N.; Yi, X.; Song, W. Shielding effectiveness and coupling characteristic of metallic enclosures with apertures under EMP. In Proceedings of the 2009 Asia-Pacific Power and Energy Engineering Conference, Wuhan, China, 28–30 March 2009.
6. Li, M.; Nuebel, J.; Drewniak, J.L.; Hubing, T.H.; DuBroff, R.E.; Van Doren, T.P. EMI Reduction from Airflow Aperture Arrays Using Dual-Perforated Screens and Loss. *IEEE Trans. EMC* **2000**, *42*, 135–141.
7. Kwon, J.H.; Hwang, J.H.; Park, H.H. Improving shielding effectiveness of enclosure with apertures using absorbers. In Proceedings of the 2019 IEEE Symposium EMC+SIPI, New Orleans, LA, USA, 22–26 July 2019.
8. Kwon, J.H.; Hyoung, C.H.; Hwang, J.H.; Park, H.H. Shielding Effectiveness Imparted by a Magnetic and Composite Absorber Placed on the Inner-Side of Aperture of Metallic Enclosure. *J. Korean Inst. Electromagn. Eng. Sci.* **2024**, *35*, 283–286. [CrossRef]
9. Kwon, J.H.; Hyoung, C.H.; Hwang, J.H.; Park, H.H. Enhancement in shielding effectiveness by electro- magnetic absorbers applied to aperture of metallic enclosure. *J. Korean Inst. Electromagn. Eng. Sci.* **2023**, *34*, 409–414. [CrossRef]
10. Kwon, J.H.; Hyoung, C.H.; Hwang, J.H.; Park, H.H. Impact of absorbers on shielding effectiveness of metallic rooms with apertures. *Electronics* **2021**, *10*, 237. [CrossRef]
11. Lee, H.J.; Kang, T.-W.; Choi, J.S.; Hong, Y.-P.; Lee, W. Shielding effectiveness of a PC case using three-axis electro-optic sensors. *J. Electromagn. Eng. Sci.* **2023**, *23*, 1–19. [CrossRef]

12. Lee, H.H.; Lim, J.-H.; Lee, J.W.; Kwon, J.H.; Hwang, J.-H.; Hyoung, C.H.; Park, H.H. Estimation of electromagnetic field penetration into concrete buildings using a theoretical approach considering external environmental factors. *J. Electromagn. Eng. Sci.* **2023**, *23*, 212–223. [CrossRef]
13. Ott, H.W. *Electromagnetic Compatibility Engineering*, 1st ed.; John Wiley & Sons, Inc.: Hoboken, NJ, USA, 2009.
14. *MIL-STD-188-125-1*; High-Altitude Electromagnetic Pulse (HEMP) Protection for Ground-Based C4I Facilities Performing Critical, Time-Urgent Missions. Department of Defense (DoD): Arlington County, VA, USA, 1998.
15. *IEEE STD 299*; IEEE Standard Method for Measuring the Effectiveness of Electromagnetic Shielding Enclosure. IEEE: Cham, Switzerland, 1997.
16. Bait-Suwailam, M.M.; Alavikia, B.; Ramahi, O.M. 'Reduction of electromagnetic radiation from apertures and enclosures using electromagnetic bandgap structures'. *IEEE Trans. Compon. Packag. Manuf. Technol.* **2014**, *4*, 929–937. [CrossRef]
17. Siah, E.S.; Sertel, K.; Volakis, J.L.; Liepa, V.V.; Wiese, R. 'Coupling studies and shielding techniques for electromagnetic penetration through apertures on complex cavities and vehicular platforms'. *IEEE Trans. EMC* **2023**, *45*, 245–257.
18. Microwave Studio (MWS), CST. 2024. Available online: www.3ds.com/products/simulia/cst-studio-suite (accessed on 27 July 2024).
19. Laird. EccosorbTM LS. 2023. Available online: https://www.laird.com (accessed on 17 August 2024).
20. GSS. Global Sourcing Supplier for You: We Are Always with You. 2022. Available online: https://www.gsskorea.com (accessed on 17 August 2024).
21. DMEGC. World Magnetic Capital. 2023. Available online: https://www.chinadmegc.com/?lang=en (accessed on 17 August 2024).

Disclaimer/Publisher's Note: The statements, opinions and data contained in all publications are solely those of the individual author(s) and contributor(s) and not of MDPI and/or the editor(s). MDPI and/or the editor(s) disclaim responsibility for any injury to people or property resulting from any ideas, methods, instructions or products referred to in the content.

electronics

MDPI

Article

Effect of Cosmic Rays on the Failure Rate of Flexible Direct Current Converter Valves in High-Altitude Environment

Liu Yang [1], Zezhao Zhang [2], Yuebin Zhou [1], Daming Wang [1], Chao Peng [2], Hong Zhang [2], Zhifeng Lei [2], Zhangang Zhang [2], Weili Fu [2,*] and Teng Ma [2,*]

1 State Key Laboratory of HVDC, Electric Power Research Institute, China Southern Power Grid, Guangzhou 510663, China
2 Science and Technology on Reliability Physics and Application of Electronic Component Laboratory, China Electronic Product Reliability and Environmental Testing Research Institute, Guangzhou 511370, China
* Correspondence: kobefwl@163.com (W.F.); mateng_ceprei@163.com (T.M.)

Abstract: Aiming at the significant needs of flexible DC converter valve applications in high-altitude areas, we investigate the effect of atmospheric neutrons on the failure rate of key core power devices in three kinds of converter valves, namely IGBTs, thyristors, and diodes. The safe working voltage boundary of the devices is obtained, and the failure rate caused by atmospheric neutrons in the real working environment of the power devices is calculated according to the results of the ground-accelerated irradiation test of atmospheric neutrons and the atmospheric neutron environment under the actual working conditions. The test results show that the failure rate of IGBTs, thyristors, and diodes caused by atmospheric neutrons is greatly affected by the blocking voltage, and the larger the blocking voltage is, the higher the failure rate of the device is. The research results can provide a basis for the design of the operating voltage of key core power devices of flexible DC converter valves and guide the evaluation of the failure rate and engineering design of the electronic system of DC ultra-high-voltage power transmission and transformation converter stations in the high-altitude environment of the Qinghai-Tibet Plateau.

Keywords: IGBT; thyristor; diode; atmospheric neutron; single-event burnout; failure rate

1. Introduction

In alignment with the national dual-carbon strategy, energy consumption in China is set to rapidly transition towards a green, low-carbon model. This shift will involve the large-scale harnessing of clean energy sources such as wind, solar, and hydroelectric power, particularly in the western regions of the country. Flexible direct current (DC) transmission technology is expected to play a crucial role in the efficient transmission of this new energy. However, the western region of China presents unique challenges due to its high altitude. Notably, there are currently no existing flexible DC transmission projects completed at altitudes above 2000 m, either domestically or internationally. Furthermore, there is a lack of data on cosmic ray atmospheric neutron tests conducted on the power devices of flexible DC converter valves [1–7]. The forthcoming results from these tests will be instrumental in advancing both domestic and international research in this field.

The flexible DC converter valve is the central component of the DC transmission system, utilizing key core power devices such as IGBT, thyristor, and diode valves to achieve AC/DC conversion. However, increasing altitude and latitude result in a rise in neutron flux, posing significant challenges. For instance, China's western region, which is generally at a higher altitude, experiences an atmospheric neutron flux over 20 times that of Guangzhou. Operating power devices at high altitudes using the same standards as in the plains may exacerbate failure problems due to cosmic rays and atmospheric neutrons.

In areas of high cosmic ray and neutron radiation, the phenomenon of degradation or failure of individual electronic components or material particles due to impact by energetic particles (e.g., neutrons, protons, etc.) is known as single-event burnout (SEB). This phenomenon is particularly important in power electronic equipment at high altitudes, where the neutron flux increases significantly, which can lead to reliability and safety issues, directly impacting the safety and stability of the DC transmission system. To address this issue, accelerated irradiation tests can be conducted in a laboratory setting, where power devices are exposed to neutron flux for periods ranging from a few hours to decades [8–11]. The results of these tests can directly inform engineering design, making it highly efficient and effective in ensuring the reliability of power devices in high-altitude environments.

In 2013, Merlin et al. used the LiCAF neutron detector for high-latitude atmospheric neutron measurements [12], and in 2022 Akin et al. conducted a study of SiC power device failure due to atmospheric neutron irradiation, and in their results, the devices did not fail significantly at neutron fluxes less than 10^{10} n/cm^2 [13]. This research addresses the critical need for versatile DC converter valve implementation in high-altitude regions by investigating the impact of atmospheric neutrons on the failure frequency of essential power core devices in three types of converter valves: IGBTs, thyristors, and diodes. Using data from atmospheric neutron ground-accelerated irradiation tests and atmospheric neutron environment assessments, this study calculates the failure rate of these devices under actual working conditions. The safe working voltage limit for each device is determined, revealing that the failure rate of IGBTs and thyristors is significantly influenced by the blocking voltage [14–16]. Specifically, the failure rate increases proportionately with an increase in blocking voltage. These findings provide a foundation for establishing optimal operating voltages for critical power devices in flexible DC converter valves. Furthermore, they offer valuable insights for assessing failure rates and guiding the engineering design of electronic systems in DC ultra-high-voltage power transmission and transformation converter stations, particularly in the high-altitude environment of the Qinghai-Tibet Plateau.

2. Materials and Methods

The irradiation tests were carried out on the Atmospheric Neutron Irradiation Spectrometer (ANIS) test bed based on the China Scattered Neutron Source (CSNS); the altitude of Dongguan, where the platform is located, is about 0–50 m above sea level. The neutron energy range of the atmospheric neutron irradiation spectrometer test platform covers meV~GeV, and the shape of the neutron energy spectrum in its samples is close to that of the natural atmospheric neutron energy spectrum, but the neutron flux is increased by a factor of 1.65×10^8 times compared with that at sea level so that it can accelerate the equivalent simulation of the effect of the natural atmospheric neutrons on the power devices. The comparison of the neutron energy spectrum of the test platform with that of the natural atmosphere is shown in Figure 1. The main parameters of the irradiated test source are shown in Table 1. A list of the main equipment for the experiment can be found in Table 2, and the experimental site plan is shown in Figure 2.

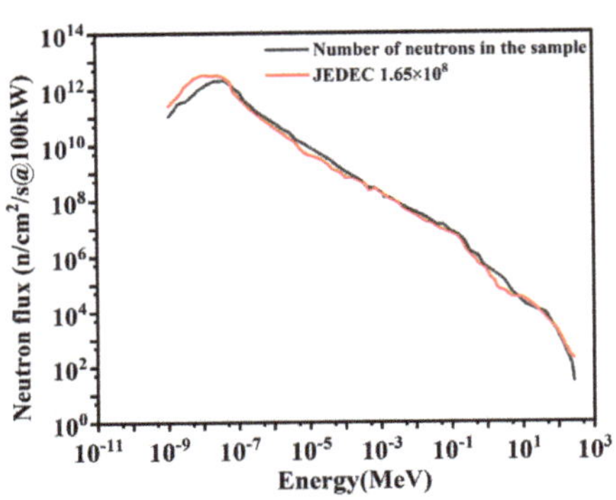

Figure 1. Comparison of the neutron energy spectrum of the irradiated test platform with the neutron energy spectrum of the natural atmosphere.

Table 1. Atmospheric neutron irradiation test platform main parameter.

Parameter Name	Parameter Index
Energy range	meV~GeV
Neutron irradiation spot area	20 cm × 15 cm
Neutron flux at sample (Energy above 10 MeV)	4.7×10^5/n/cm^2/s

Table 2. List of major equipment and its model.

Equipment Name	Equipment Model
Atmospheric Neutron Irradiation Spectrometer	ANIS
High-voltage DC power supplies	DSP
High- and low-temperature box	NT125-55AS

(a) (b) (c)

Figure 2. Three devices' experimental field diagram: (**a**) IGBT chip; (**b**) diode chip; (**c**) thyristor chip.

In the process of the atmospheric neutron acceleration and irradiation test, it is necessary to bias the power device in the high-voltage blocking state and monitor its off-state current in real-time, and its test principle is shown in Figure 3. In the figure, from left to right, the scattering neutron source, the sample to be tested and the fixture, the multi-channel test platform, and the high-voltage DC power supply are shown in order. During the irradiation test, the samples and fixtures to be tested are placed in the irradiation room, and the test equipment (including the high-voltage DC power supply, the multi-channel test platform, the computer, etc.) and the testers are in the test room. The high-voltage power supply is connected to the multi-channel test platform and then connected to the sample to be tested through a cable of more than 30 m to supply power to it.

To enhance experimental efficiency, accelerated testing involves irradiating multiple devices simultaneously. During the test, all devices are connected in parallel and powered by a high-voltage source. If a device experiences a single-event burnout, it loses its blocking capability and exhibits a short-circuit characteristic. This failure can prevent the other parallel-connected devices from being properly energized. Therefore, it is necessary to disconnect the failed device branch from the power supply. In order to achieve the above purpose, a multi-channel test platform is used to realize parallel testing of multiple samples. The multi-channel test platform mainly consists of a control circuit board, current sensors, and relays. Each sample to be tested is connected to the same high-voltage power supply through the sensor and relay and then irradiated at the same time [17–19]. When a sample fails, the sensor detects a change in the corresponding current and sends a signal to the control circuit, which then controls the relay to disconnect from the high-voltage power supply, thus ensuring that the voltage is only applied to the non-failed sample.

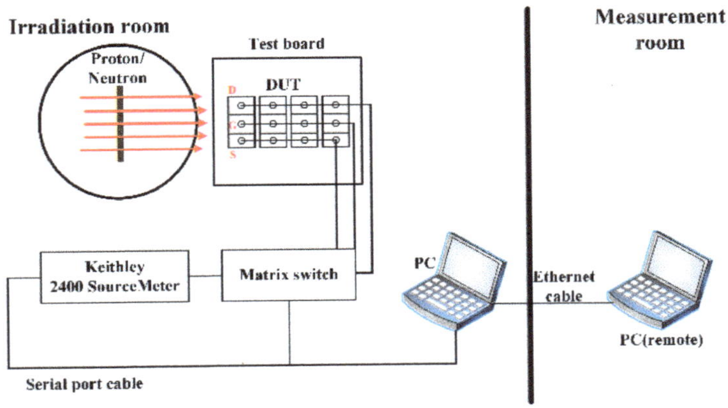

Figure 3. Diagram of the experimental setup for the atmospheric neutron irradiation test, where the irradiation room is on the left and the test room is on the right.

During irradiation, a sample is considered to have failed if there is an instantaneous surge in off-state current and a loss of blocking capability. One failure event is recorded if a sudden surge in off-state current of 100 mA is detected during irradiation. After calculating the neutron flux at 29.79° N, 97.71° E, and 4000 m above sea level, the neutron flux above 10 MeV that may lead to the failure of the power device is 116 $n/cm^2/h$. According to the requirement that the failure rate of the thyristor device is not higher than 50 FIT, i.e., 50 samples are failing in 1×10^9 device hours, 5 chips are selected and irradiated at the same time. Galvanic irradiation, and then the total irradiation of each test chip, is approximately as follows:

$$116\ n/cm^2/h \times 1 \times 10^9\ h/(5 \times 50) = 4.64 \times 10^8\ n/cm^2$$

Therefore, when the thyristor chip is used to carry out the test, the maximum neutron injection during the test is set at 4.64×10^8 n/cm^2, and the irradiation test can be stopped even if there is no device failure when the cumulative neutron injection during the test reaches 4.64×10^8 n/cm^2. The atmospheric neutron accelerated irradiation test is carried out under selected temperature and voltage conditions, and the test obtains the failure rate at different voltages and temperatures. Since the average temperature in Tibet is 3–17 °C, we conducted the experiment at a lower temperature of 5 °C to restore the temperature of the real application scenario; the irradiation test is carried out at a minimum of 5 different voltages for each sample, and the neutron irradiation injection and bias voltage chosen for the experiment are the result of a combination of experimental conditions and recommendations from the literature [13].

A matching fixture is designed for the test sample. The fixtures are used to hold the samples to be tested and to ensure that the neutrons are incident vertically on the samples. The fixture is also used to lead out the charging port of each sample. Among them, the IGBT chip and the diode chip share a common set of fixtures. The cross-sectional size of all the samples mounted in the fixture does not exceed the size of the neutron beam spot for the test (20 cm × 15 cm). Through laser positioning, it is ensured that the neutron beam spot can cover all the samples to be tested. Throughout the test, the samples to be tested were placed in a high- and low-temperature chamber, which was set at a constant temperature of 5 °C. The chamber had a 40 cm × 40 cm space for the sample to be tested. A 40 cm × 40 cm glass viewing port is left in the chamber, through which the neutrons irradiate the sample to be tested.

Record the number of samples r for which failure occurred during the irradiation test and record the total effective sample injection T_{SUM} (neutron injection with energy above 10 MeV):

$$T_{SUM} = \sum n_i \quad (1)$$

where i is the ith test sample, n_i is the neutron injection at the time of failure of the ith sample, or n_i is the cumulative neutron injection at the time of stopping the beam if no failure of this sample occurs. Thus, the accelerated test failure rate λ_{ACC} is as follows:

$$\lambda_{ACC} = r/T_{SUM} \quad (2)$$

Based on the number of sample failures r recorded in the accelerated irradiation test, the corresponding confidence interval for the failure rate can be calculated as follows:

$$(2T_{SUM})^{-1}X^2_{\alpha/2}(2r) < \lambda_{ACC} < (2T_{SUM})^{-1}X^2_{1-\alpha/2}(2r) \quad (3)$$

where X^2 is the chi-square function with a degree of freedom of $2r$; T_{SUM} is the total neutron injection of the effective sample. The confidence intervals are all calculated in this report according to the 95% confidence level, which corresponds to the 95% confidence level of α = 5%. The actual device failure rate λ_{ACC} is obtained by multiplying the failure rate λ_{ACR} obtained in the test by the average neutron flux Φ_n (energy above 10 MeV) in the actual environment:

$$\lambda_{ACR} = \lambda_{ACC} \times \Phi_n \quad (4)$$

Based on the results of the accelerated irradiation test, the device failure rates corresponding to 4000 m above sea level, 2000 m above sea level, and sea level are calculated in this work. In particular, the natural atmospheric neutron flux (Φn) above 10 MeV varies significantly with altitude. At an altitude of 4000 m, the flux is calculated to be 116 $n/cm^2/s$. This value decreases to 32 $n/cm^2/s$ at an altitude of 2000 m. At sea level, the neutron flux further diminishes to 6.2 $n/cm^2/s$. These calculations highlight the inverse relationship between altitude and neutron flux, with higher altitudes experiencing greater neutron activity.

3. Results and Discussions

Figure 3, showing the results for the IGBT chips, depicts the irradiation tests that were carried out under five voltage conditions, 2000 V, 2050 V, 2100 V, 2200 V, and 2300 V, respectively, at 5 °C. When the IGBT chips were operated at 2000 V, no failure was observed in the tests, and the failure rate at 4000 m above sea level was less than 0.96 FIT; when they were operated at 2050 V, the failure rate of a single chip was 1.96 FIT at 4000 m above sea level. The test results of the IGBT chip and the failure rate calculation results are shown in Table 3, and the atmospheric neutron failure rate of the IGBT single chip at different altitudes is shown in Figure 4.

Table 3. Statistics of IGBT chip failure due to atmospheric neutron irradiation at different bias voltages.

No.	Voltage (V)	Number of Samples	Number of Failures	Total Effective Neutron Injection (n/cm^2)
1	2000	30	0	1.21×10^{11}
2	2050	30	2	1.18×10^{11}
3	2100	30	3	1.13×10^{11}
4	2200	30	11	7.83×10^{10}
5	2300	37	23	3.54×10^{10}

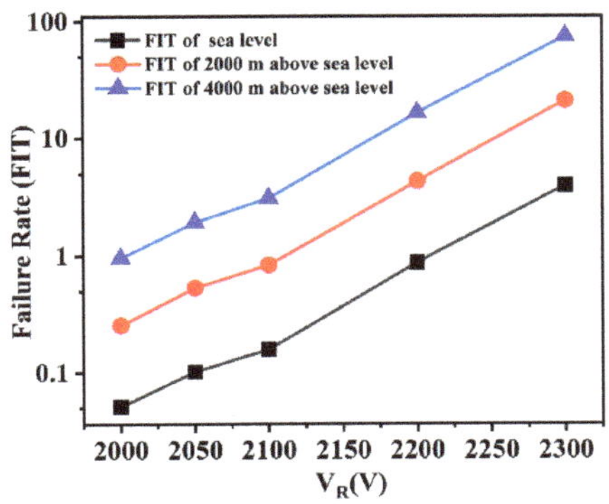

Figure 4. Curve of IGBT chips' failure rate versus bias voltage due to atmospheric neutron irradiation at different altitudes (temperature = 5 °C).

According to the cosmic ray failure rate model for power devices proposed by Zeller et al. [9], the SEB failure rate is exponentially related to the applied voltage. The parameters a and b in the failure model require experimental testing, as they vary across different device models.

$$\lambda(T, V_{DC}) = a_{(T)} e^{-\frac{b_{(T)}}{V_{DC}}} \tag{5}$$

In this context, λ represents the failure rate, and V_{DC} is the applied direct current voltage. The SEB failure caused by cosmic rays is inherently random. For a given voltage, temperature, and altitude, the failure rate remains constant throughout the device's lifecycle. High-speed proton irradiation tests are used to simulate atmospheric cosmic ray exposure, allowing for a rapid determination of the thyristor SEB failure rate. To prevent unexpected results and enhance the accuracy of the failure rate experiments, 3–5 samples are tested under each set of conditions. The average failure rate is calculated to represent the failure rate of the wafer or device under those conditions. The formula for calculating the average failure rate *FR* is as follows:

$$FR = \frac{r}{(m-r) \times AF + \sum_{i=1}^{r} AFi} \times 3.6 \times 10^{12} FIT \tag{6}$$

In this context, *r* represents the number of failures observed under a specific irradiation condition in the experiment, while *m* denotes the total number of device samples under the same condition; both *r* and *m* are measured in units. *AF* is the equivalent experimental time, calculated as the ratio of the total fluence of non-failed devices to the atmospheric neutron flux in the actual application environment. *AFi* is the ratio of the total fluence for the ith device during the experiment to the atmospheric neutron flux in the actual application environment, measured in seconds. The neutron irradiation test results were processed using the formula for calculating the average failure rate of devices, yielding the failure rate of the IGBT chip under neutron irradiation, as shown in Table 4. In the experimental process, due to the device failure at a lower bias voltage, a single experiment time is too long; with regard to the device residual radiation hazards, taking into account the safety of experimental personnel and the reliability of the experimental equipment, when the bias voltage of the device does not fail, we will consider increasing the bias span for the experiment, making it easy to find the voltage value of the failed device. At the same time, in the vicinity of the voltage in which failure of the device occurred, we will then carry out an adjustment of the voltage up and down. At the same time, in the vicinity of the device failure voltage value, we will then conduct the experiment within the time interval, that is, consider increasing the bias span for the experiment, which is the reason why our test voltage is not exact according to the same step [20].

Table 4. Statistical results of chip failure rate of IGBT chips due to atmospheric neutron irradiation with altitude at different bias voltages.

	Voltage (V)	Failure Rate at 4000 m (FIT)	4000 m Failure Rate Confidence Interval (FIT, 95% Confidence)	Failure Rate at 2000 m (FIT)	2000 m Failure Rate Confidence Interval (FIT, 95% Confidence)	Failure Rate at Sea Level (FIT)	Sea Level Failure Rate Confidence Interval (FIT, 95% Confidence)
1	2000	$<9.60 \times 10^{-1}$	/	$<2.65 \times 10^{-1}$	/	$<5.13 \times 10^{-2}$	/
2	2050	1.96	2.38×10^{-1}, 5.46	5.41×10^{-1}	6.55×10^{-2}, 1.51	1.05×10^{-1}	1.27×10^{-2}, 2.92×10^{-1}
3	2100	3.08	6.35×10^{-1}, 7.41	8.49×10^{-1}	1.75×10^{-1}, 2.04	1.64×10^{-1}	3.39×10^{-2}, 3.96×10^{-1}
4	2200	16.3	8.13, 27.2	4.49	2.24, 7.51	8.71×10^{-1}	4.35×10^{-1}, 1.46
5	2300	75.4	47.8, 109	20.8	13.2, 30.1	4.03	2.56, 5.84

For the built-in diode chip of 4500 V/3000 IGBT, ten voltage conditions of 2100 V, 2150 V, 2200 V, 2250 V, 2300 V, 2700 V, 2800 V, 2900 V, 3000 V, and 3100 V were selected at 5 °C for the irradiation test. No failure was observed when the diode chip was operated below 2150 V. The failure rate of a single chip at 2200 V, corresponding to an altitude of 4000 m, was 8.58 FIT. At the same voltage of 2200 V, the failure rate of the diode chip was half that of the IGBT chip. The test results of the diode chip are recorded, and the failure rate is calculated as shown in Table 5. The calculated results of atmospheric neutron failure rates of built-in diode chips at different voltages are shown in Table 6. The curves of the atmospheric neutron failure rate of the diode single chip with different voltages at different altitudes are shown in Figure 5.

Table 5. Statistics of diode chips' failure due to atmospheric neutron irradiation at different bias voltages.

No.	Voltage (V)	Number of Samples	Number of Failures	Total Effective Neutron Injection (n/cm^2)
1	2100	15	0	5.94×10^{10}
2	2150	15	0	5.92×10^{10}
3	2200	15	3	4.05×10^{10}
4	2250	15	4	1.44×10^{10}
5	2300	15	10	2.78×10^{10}
6	2700	15	12	5.46×10^{9}
7	2800	15	13	1.99×10^{9}
8	2900	15	14	8.32×10^{8}
9	3000	15	14	3.17×10^{8}
10	3100	15	13	1.33×10^{8}

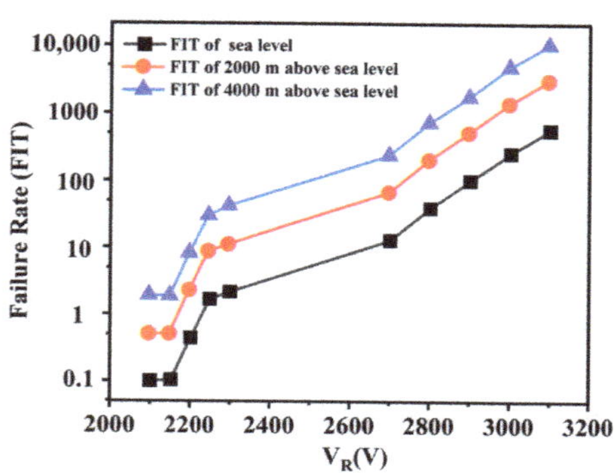

Figure 5. Curve of diode chips' failure rate versus bias voltage due to atmospheric neutron irradiation at different altitudes (temperature = 5 °C).

Table 6. Statistical results of chip failure rate of diode chips due to atmospheric neutron irradiation with altitude at different bias voltages.

	Voltage (V)	Failure Rate at 4000 m (FIT)	4000 m Failure Rate Confidence Interval (FIT, 95% Confidence)	Failure Rate at 2000 m (FIT)	2000 m Failure Rate Confidence Interval (FIT, 95% Confidence)	Failure Rate at Sea Level (FIT)	Sea Level Failure Rate Confidence Interval (FIT, 95% Confidence)
1	2100	<1.95	/	$<5.39 \times 10^{-1}$	/	$<1.04 \times 10^{-1}$	/
2	2150	<1.96	/	$<5.40 \times 10^{-1}$	/	$<1.05 \times 10^{-1}$	/
3	2200	8.58	1.77, 20.7	2.37	4.88×10^{-1}, 5.70	4.59×10^{-1}	9.46×10^{-2}, 1.10
4	2250	32.3	8.79, 70.7	8.90	2.43, 19.5	1.72	4.70×10^{-1}, 3.78
5	2300	41.7	20, 71.2	11.5	5.51, 19.6	2.23	1.07, 3.81
6	2700	255	132, 418	70.3	36.3, 115	13.6	7.04, 22.3
7	2800	759	404, 1220	2.09×10^2	1.12×10^2, 3.38×10^2	40.6	21.6, 65.4
8	2900	1.95×10^3	1.07×10^3, 3.10×10^3	5.38×10^2	2.94×10^2, 8.55×10^2	1.04×10^2	57.0, 1.66×10^2
9	3000	5.12×10^3	3.90×10^3, 8.14×10^3	1.41×10^3	1.08×10^3, 2.24×10^3	2.74×10^2	2.09×10^2, 4.35×10^2
10	3100	1.14×10^4	6.05×10^3, 1.83×10^4	3.14×10^3	1.67×10^3, 5.06×10^3	6.07×10^2	3.23×10^2, 9.80×10^2

For the thyristor chip, six voltage conditions, 2000 V, 2100 V, 2150 V, 2250 V, 2300 V, and 2400 V, were selected at 5 °C for the irradiation test. No failure was observed when the thyristor chip was operated at 2000 V. The failure rate of a single chip at 2100 V, corresponding to an altitude of 4000 m above sea level, was 76.4 FIT. The test results of the thyristor chip and the calculation of the failure rate are shown in Table 7, and the calculated results of atmospheric neutron failure rates of bypass thyristor chips at different voltages are shown in Table 8. The atmospheric neutron failure rate of the thyristor chip at different altitudes is shown in Figure 6.

Table 7. Statistics of thyristor chips' failure due to atmospheric neutron irradiation at different bias voltages.

No.	Voltage (V)	Number of Samples	Number of Failures	Total Effective Neutron Injection (n/cm²)
1	2000	5	0	1.01×10^{10}
2	2100	5	3	4.56×10^9
3	2150	5	5	3.71×10^9
4	2250	5	4	2.27×10^9
5	2300	5	5	1.96×10^7
6	2400	5	3	1.81×10^6

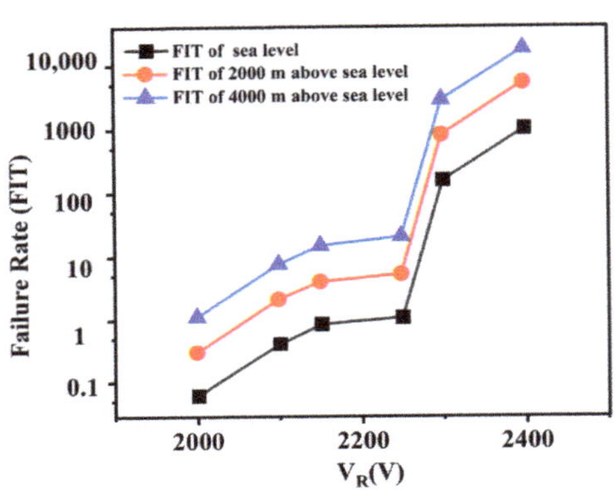

Figure 6. Curve of thyristor chips' failure rate versus bias voltage due to atmospheric neutron irradiation at different altitudes (temperature = 5 °C).

Table 8. Statistical results of chip failure rate of diode chips due to atmospheric neutron irradiation with altitude at different bias voltages.

	Voltage (V)	Failure Rate at 4000 m (FIT)	4000 m Failure Rate Confidence Interval (FIT, 95% Confidence)	Failure Rate at 2000 m (FIT)	2000 m Failure Rate Confidence Interval (FIT, 95% Confidence)	Failure Rate at Sea Level (FIT)	Sea Level Failure Rate Confidence Interval (FIT, 95% Confidence)
1	2000	<11.5	/	<3.18	/	$<6.16 \times 10^{-1}$	/
2	2100	76.4	$15.7, 1.84 \times 10^2$	21.1	4.34, 50.7	4.08	8.42×10^{-1}, 9.83
3	2150	1.56×10^2	$50.8, 3.20 \times 10^2$	43.2	14.0, 88.4	8.36	2.72, 17.1
4	2250	2.04×10^2	$55.7, 4.48 \times 10^2$	56.4	$15.4, 1.24 \times 10^2$	10.9	2.98, 23.9
5	2300	2.95×10^4	9.59×10^3, 6.05×10^4	8.14×10^3	2.64×10^3, 1.67×10^4	1.58×10^3	5.12×10^2, 3.23×10^3
6	2400	1.92×10^5	1.02×10^6, 1.19×10^7	5.31×10^4	2.81×10^5, 3.28×10^6	1.03×10^4	5.44×10^4, 6.35×10^5

When high-energy neutrons (≥10 MeV) strike power devices such as IGBTs, diodes, and thyristors, they cause ionizing recoils with silicon atoms in the substrate material. This interaction generates a large number of electrons and holes locally within the device, disrupting the electric field in the space charge region and leading to single-event burnout (SEB). This SEB, induced by high-energy neutrons, is a primary cause of device failure [21–23]. We conducted a failure analysis on IGBTs that experienced SEB, and the results are shown in Figure 7. Figure 7a presents the OBIRCH analysis of the damaged area, while Figure 7b shows the SEM morphology of the damaged region.

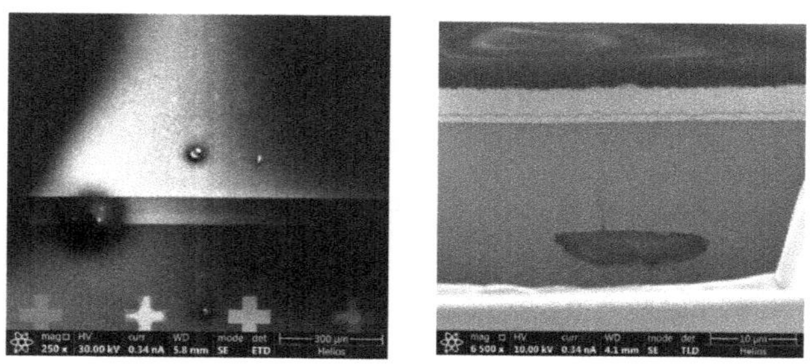

(a) (b)

Figure 7. Failure analysis of the IGBT with SEB. (**a**) OBIRCH analysis of IGBT. The "bright spot" area represents the damaged region. (**b**) SEM diagram of the damaged region after stripping the metal layer.

Based on the results of the accelerated irradiation test and the atmospheric neutron environment under the actual working conditions, the failure rate caused by atmospheric neutrons under the real working environment of the power devices is calculated. The test results show that the failure rate of IGBTs and thyristors caused by atmospheric neutrons is greatly affected by the blocking voltage [24]. The larger the blocking voltage, the higher the failure rate. For IGBT chips, when working at 2000 V, the failure rate at 4000 m above sea level is less than 0.96 FIT; when working at 2050 V, the failure rate of a single chip at 4000 m above sea level is 1.96 FIT; for a diode chip, when working at 2150 V, the failure rate at 4000 m above sea level is less than 1.96 FIT; when working at 2200 V, the failure rate at 2200 V is less than 1.96 FIT. For thyristor chips, the failure rate at 2000 V is less than 11.5 FIT at 4000 m. At 2100 V, the failure rate at 4000 m is 76.4 FIT.

4. Conclusions

This study simulates the single-event burnout (SEB) effects on power devices such as IGBTs, diodes, and thyristors in flexible DC converter valves caused by high-energy neutrons in the atmospheric environment using a neutron irradiation spectrometer. Accelerated tests were conducted using an equivalent total particle injection method, confirming the presence of SEB under neutron irradiation. The failure was characterized by a significant increase in off-state current, leading to the loss of the device's blocking capability. Additionally, a method for calculating the average SEB failure rate of power devices like IGBTs, diodes, and thyristors was proposed. Results showed that the failure rate of these devices increases with applied voltage and is positively correlated with altitude, due to the increase in neutron flux at higher altitudes. These findings provide a reference for designing the operating voltage of thyristors in flexible DC converter valves, avoiding excessive redundancy and enhancing the reliability and cost-effectiveness of DC transmission projects.

Author Contributions: Conceptualization, W.F.; Methodology, Y.Z.; Formal analysis, D.W.; Data curation, C.P. and Z.Z. (Zhangang Zhang); Writing—original draft, L.Y.; Writing—review & editing, T.M.; Visualization, H.Z.; Project administration, Z.Z. (Zezhao Zhang) and Z.L. All authors have read and agreed to the published version of the manuscript.

Funding: This work was supported by the State Key Laboratory of HVDC (No. SKLHVDC-2022-KF-08).

Data Availability Statement: All data are available from the corresponding authors.

Conflicts of Interest: There are no conflicts of interest in this article.

References

1. Liu, Q.; Tai, S.; Lu, W.; Sun, J.; Lv, T.; Liu, C.; Sun, Y.; Lv, J.; Liu, W.; Sun, T.; et al. Design of pure silica-based photonic crystal fiber for supporting 114 OAM modes transmission. *J. Opt.* **2021**, *23*, 095701. [CrossRef]
2. Lu, F.; Wright, R.; Lu, P.; Cvetic, P.C.; Ohodnicki, P.R. Distributed fiber optic pH sensors using sol-gel silica based sensitive materials. *Sens. Actuators B Chem.* **2021**, *340*, 129853. [CrossRef]
3. Saiding, Q.; Cui, W. Functional nanoparticles in electrospun fibers for biomedical applications. *Nano Sel.* **2022**, *3*, 999–1011. [CrossRef]
4. Girard, S.; Morana, A.; Ladaci, A.; Robin, T.; Mescia, L.; Bonnefois, J.-J.; Boutillier, M.; Mekki, J.; Paveau, A.; Cadier, B.; et al. Recent advances in radiation-hardened fiber-based technologies for space applications. *J. Opt.* **2018**, *20*, 093001. [CrossRef]
5. Girard, S.; Kuhnhenn, J.; Gusarov, A.; Brichard, B.; Uffelen, M.V.; Ouerdane, Y.; Boukenter, A.; Marcandella, C. Radiation Effects on Silica-Based Optical Fibers: Recent Advances and Future Challenges. *IEEE Trans. Nucl. Sci.* **2013**, *60*, 2015–2036. [CrossRef]
6. Barrera, D.; Madrigal, J.; Delepine-Lesoille, S.; Sales, S. Multicore optical fiber shape sensors suitable for use under gamma radiation. *Opt. Express* **2019**, *27*, 29026–29033. [CrossRef] [PubMed]
7. Girard, S.; Alessi, A.; Richard, N.; Martin-Samos, L.; De Michele, V.; Giacomazzi, L.; Agnello, S.; Francesca, D.D.; Morana, A.; Winkler, B.; et al. Overview of radiation induced point defects in silica-based optical fibers. *Rev. Phys.* **2019**, *4*, 100032. [CrossRef]
8. Zeller, H.R. Cosmic ray induced breakdown in high voltage semiconductor devices, microscopic model and phenomenological lifetime prediction. In Proceedings of the 6th International Symposium on Power Semiconductor Devices and ICs, Davos, Switzerland, 31 May–2 June 1994; pp. 339–340.
9. Zeller, H.R. Cosmic ray induced failures in high power semiconductor devices. *Solid-State Electron.* **1995**, *38*, 2041–2046. [CrossRef]
10. Prado, R.A.; Gajo, C.N.L. Power semiconductor failures due to cosmic rays. In Proceedings of the 2017 Brazilian Power Electronics Conference (COBEP), Juiz de Fora, Brazil, 19–22 November 2017; pp. 1–6.
11. Matsuda, H.; Fujiwara, T.; Hiyoshi, M.; Nishitani, K.; Kuwako, A.; Ikehara, T. Analysis of GTO failure mode during DC voltage blocking. In Proceedings of the 6th International Symposium on Power Semiconductor Devices and ICs, Davos, Switzerland, 31 May–2 June 1994; pp. 221–225.
12. Kole, M.; Fukazawa, Y.; Fukuda, K.; Ishizu, S.; Jackson, M.; Kamae, T.; Kawaguchi, N.; Kawano, T.; Kiss, M.; Moretti, E.; et al. A balloon-borne measurement of high latitude atmospheric neutrons using a licaf neutron detector. In Proceedings of the 2013 IEEE Nuclear Science Symposium and Medical Imaging Conference (2013 NSS/MIC), Seoul, Republic of Korea, 27 October–2 November 2013; pp. 1–8. [CrossRef]
13. Akturk, A.; Wilkins, R.; Gunthoti, K.; Wender, S.A.; Goldsman, N. Energy Dependence of Atmospheric Neutron-Induced Failures in Silicon Carbide Power Devices. *IEEE Trans. Nucl. Sci.* **2022**, *69*, 900–907. [CrossRef]
14. Zhang, Z.; Lei, Z.; Shi, Q.; Yue, L.; Huang, Y.; En, Y. Neutron Radiation Environment and Result Single Event Effects Prediction in Near Space. *J. Space Sci.* **2018**, *38*, 502–507. (In Chinese)
15. Cai, M.; Zhang, Z.; Feng, G.; Zhu, L.; Han, J. Study of Near Space Neutron Environment and Its Effects on Electronic Device. *Equip. Environ. Eng.* **2007**, *4*, 23–29. (In Chinese)
16. Wang, X.; Zhang, F.; Chen, W.; Guo, X.; Ding, L.; Luo, Y. Application and evaluation of Chinese spallation neutron source in single-event effects testing. *Acta Phys. Sin.* **2019**, *68*, 38–47. (In Chinese) [CrossRef]

17. Regnier, E.; Flammer, I.; Girard, S.; Gooijer, F.; Achten, F.; Kuyt, G. Low-Dose Radiation-Induced Attenuation at InfraRed Wavelengths for P-Doped, Ge-Doped and Pure Silica-Core Optical Fibres. *IEEE Trans. Nucl. Sci.* **2007**, *54*, 1115–1119. [CrossRef]
18. Girard, S.; Keurinck, J.; Boukenter, A.; Meunier, J.P.; Ouerdane, Y.; Azaıs, B.; Charre, P.; Vié, M. Gamma-rays and pulsed X-ray radiation responses of nitrogen-, germanium-doped and pure silica core optical fibers. *Nucl. Instrum. Methods Phys. Res. Sect. B Beam Interact. Mater. At.* **2004**, *215*, 187–195. [CrossRef]
19. Girard, S.; Marcandella, C.; Alessi, A.; Boukenter, A.; Ouerdane, Y.; Richard, N.; Paillet, P.; Gaillardin, M.; Raine, M. Transient Radiation Responses of Optical Fibers: Influence of MCVD Process Parameters. *IEEE Trans. Nucl. Sci.* **2012**, *59*, 2894–2901. [CrossRef]
20. Peng, C.; Lei, Z.; Zhang, Z.; Yang, S.; Lai, P.; Lu, G. Measurement and Simulation of Ground-level Atmospheric Neutron Energy Spectra in Typical Regions of China. *High Power Laser Part. Beams* **2023**, *35*, 153–158. (In Chinese) [CrossRef]
21. Li Vecchi, G.; Di Francesca, D.; Sabatier, C.; Girard, S.; Alessi, A.; Guttilla, A.; Robin, T.; Kadi, Y.; Brugger, M. Infrared radiation Induced attenuation of radiation sensitive optical fibers: Influence of temperature and modal propagation. *Opt. Fiber Technol.* **2020**, *55*, 102166. [CrossRef]
22. Faustov, A.V.; Gusarov, A.; Wuilpart, M.; Fotiadi, A.A.; Liokumovich, L.B.; Zolotovskiy, I.O.; Tomashuk, A.L.; Schoutheete, T.d.; Mégret, P. Comparison of Gamma-Radiation Induced Attenuation in Al-Doped, P-Doped and Ge-Doped Fibres for Dosimetry. *IEEE Trans. Nucl. Sci.* **2013**, *60*, 2511–2517. [CrossRef]
23. Francesca, D.D.; Boukenter, A.; Agnello, S.; Girard, S.; Alessi, A.; Paillet, P.; Marcandella, C.; Richard, N.; Gelardi, F.M.; Ouerdane, Y. X-ray irradiation effects on fluorine-doped germanosilicate optical fibers. *Opt. Mater. Express* **2014**, *4*, 1683–1695. [CrossRef]
24. Smietana, M.; Bock, W.J.; Mikulic, P.; Chen, J. Tuned Pressure Sensitivity of Dual Resonant Long-Period Gratings Written in Boron Co-Doped Optical Fiber. *J. Light. Technol.* **2012**, *30*, 1080–1084. [CrossRef]

Disclaimer/Publisher's Note: The statements, opinions and data contained in all publications are solely those of the individual author(s) and contributor(s) and not of MDPI and/or the editor(s). MDPI and/or the editor(s) disclaim responsibility for any injury to people or property resulting from any ideas, methods, instructions or products referred to in the content.

electronics

MDPI

Article

Regression Analysis for Predicting the Magnetic Field Shielding Effectiveness of Ferrite Sheets

Hyun Ho Park [1,*], Heehyuk Lee [2] and Deuk-Kyu Hwang [2]

1 School of Electrical and Electronic Engineering, The University of Suwon, Hwaseong 18323, Republic of Korea
2 Metallic & Electrical Materials Team, Hyundai MOBIS Co., Ltd., Yongin 16891, Republic of Korea; heehyuk2@mobis.com (H.L.); hdk@mobis.com (D.-K.H.)
* Correspondence: hhpark@suwon.ac.kr

Abstract: In this paper, a method to predict near-field magnetic shielding effectiveness (NSE) of ferrite sheets is proposed by measuring their relative permeability. The NSE prediction for ferrite sheets is developed using eight regression models based on higher-order terms of permeability, extracted through Minitab's regression analysis using data from the measured NSE and relative permeabilities of the ferrite sheets. To analyze the accuracy of the predicted NSE in comparison to the measured NSE, the mean square error (MSE) was computed. As a result, the extracted regression models enable fast and accurate NSE predictions for ferrite sheets up to 100 MHz, achieving an MSE of less than 1.0, in contrast to numerical simulation methods that require several hours.

Keywords: magnetic ferrite sheet; magnetic shielding effectiveness; relative permeability; regression model; mean square error

1. Introduction

In modern electronic devices, high clock speeds, densely packed PCB layouts, and the proximity of high-speed circuits exacerbate electromagnetic interference (EMI) issues. Furthermore, the extensive use of wireless communication and sophisticated power management systems adds to the challenge. As concerns about EMI grow, the importance of effective shielding and noise suppression techniques is also increasing.

Magnetic sheets, typically made from flexible magnetic materials like ferrite or metal alloy composites, are designed to absorb and dissipate unwanted electromagnetic noise that can interfere with electronic circuits and components as shown in Figure 1. Furthermore, when the coupling between noisy components and susceptible components increases due to the metallic boxes or enclosures that package them, attaching magnetic sheets to the metallic surfaces can help reduce noise coupling caused by the induced currents on the inner conducting plane, as depicted in Figure 2. Therefore, in an increasingly noisy electromagnetic environment, magnetic sheets are indispensable for maintaining the performance and integrity of electronic devices.

In particular, noise suppression sheets (NSSs) are commonly employed for reducing and suppressing noise coupling and propagation at the GHz frequency range in modern electronic devices. Numerous studies have been conducted on NSSs [1–5]. In addition to their use in the GHz high-frequency range, magnetic sheets are increasingly being employed to reduce low-frequency noise in electric vehicles [6,7]. The noise reduction mechanisms of magnetic sheets differ between high-frequency and low-frequency ranges. In high-frequency ranges, noise is absorbed due to the loss characteristics associated with

https://doi.org/10.3390/electronics14020310

the imaginary component of the permeability of the magnetic material. In contrast, in low-frequency ranges, the real component of the permeability causes the magnetic field to concentrate within the magnetic material, thereby converting or guiding the magnetic field path to reduce noise coupling between the noise source and victim circuits.

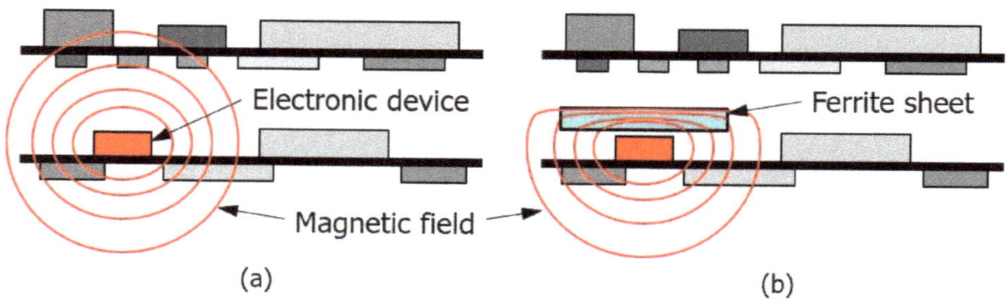

Figure 1. Usage of magnetic sheets for a reduction in EMI in densely packed electronic devices. (**a**) Without magnetic sheet. (**b**) With magnetic sheet.

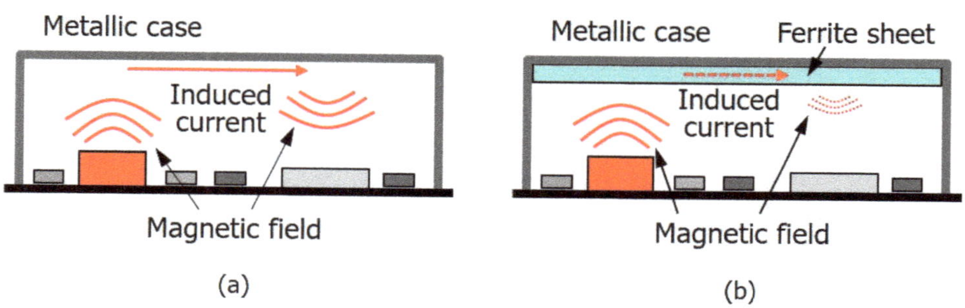

Figure 2. Usage of magnetic sheets for suppressing noise coupling of electronic devices in metallic cases due to the induced current on inner metal surface. (**a**) Without magnetic sheet. (**b**) With magnetic sheet.

During the design phase of electronic devices, accurately predicting the near-field magnetic shielding properties of magnetic sheets is essential for selecting the most suitable materials to mitigate noise coupling. However, conventional approaches that rely on numerical simulations, such as finite element analysis (FEA) or other complex computational methods, have several limitations. These simulations often demand computational time and resources due to the complexity of the magnetic materials and the need to model intricate device geometries accurately. Additionally, they require specialized expertise to set up and interpret the results effectively, making the process labor-intensive and potentially impractical for quick design iterations. As a result, there is a need for more efficient prediction methods that can provide reliable results with less effort and time investment.

In this paper, we propose a method to predict the near-field magnetic shielding effectiveness (NSE) of the ferrite-based magnetic materials up to 100 MHz by measuring their real permeability using regression analysis. To extract the eight regression models, complex relative permeability and NSE were measured using commercial ferrite sheets. Additionally, the differences or errors between the predicted NSE from the extracted regression models and the measured NSE were analyzed using an error assessment indicator: mean square error (MSE). The results confirmed that the proposed regression models provide sufficiently accurate predictions of NSE in the low-frequency range up to 100 MHz. The proposed method achieves comparable accuracy with negligible computational time when compared to numerical simulations.

2. Methodology

The procedure of regression analysis consists of three phases for predicting the NSE of ferrite sheets and for verifying the accuracy of the predicted results, as illustrated in Figure 3. In phase 1, regression models were developed using the measured permeability and NSE data of five commercial ferrite sheets (Group #1). The regression analysis was performed using Minitab [8]. In phase 2, the NSE was predicted using the measured permeability data of another two commercial ferrite sheets (Group #2) as input for the extracted regression models. In phase 3, the NSE of the ferrite sheets in Group #2 was measured, and the accuracy was verified by comparing them with the predicted NSE. The predicted NSE of the ferrite sheets in Group #1 used for extracting the regression models was also compared with the measured NSE.

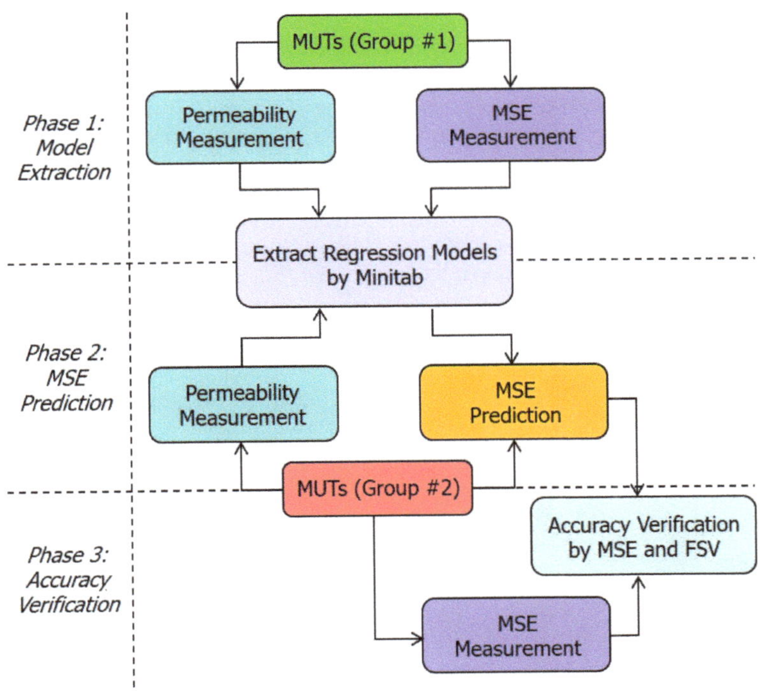

Figure 3. Flow chart of three-phase procedure for regression analysis and prediction.

The prediction of NSE is made by the regression models, which were extracted through the measured frequency-dependent permeabilities and the NSE of five commercial ferrite sheets (Group #1). For NSE prediction, the measured permeability is required not only for the ferrite sheet in Group #1 but also for the ferrite sheets in Group #2. The complex relative permeability of the ferrite sheets was measured from 1 kHz to 100 MHz using the 16454A magnetic material test fixture provided by Keysight [9]. A material under test (MUT) with a toroidal core shape and a thickness below 3 mm was coiled with a wire. Its relative permeability was calculated by measuring the impedance of the MUT based on the relationship between inductance and permeability. Table 1 lists the measured real relative permeability at 1 MHz and the thickness of the commercial ferrite sheets used in the analysis.

Table 1. Real relative permeability at 1 MHz and thickness of ferrite sheets used in the analysis.

Name of MUTs		Real Relative Permeability at 1 MHz	Thickness [1]
Group #1	R1	143	0.4
	R2	166	0.4
	R3	273	0.2
	R4	316	0.4
	R5	734	0.2
Group #2	P1	152	0.4
	P2	324	0.3

[1] The unit is mm.

The NSE was measured using a microstrip line and loop probe [10–12]. Figure 4 shows a fabricated test fixture for magnetic shielding measurement of a thin magnetic sheet, which consists of a test board with a 50 Ω microstrip line and a loop probe. The size of the test board is 140 mm × 140 mm and the thickness is 2.0 mm with FR4 substrate (ε_r = 4.3). The length and width of the microstrip line are 60 mm and 3.78 mm, respectively. To capture the leakage near-field passing through a magnetic sheet, a magnetic loop probe (RF-R 400-1 [13]) with a diameter is 24 mm is employed. The distance (h) between the loop probe and microstrip line is set to be 1 mm. The size of ferrite sheets is 140 × 140 mm, which is the same as that of the test board and is sufficiently larger than that of the loop probe. The ferrite sheets are mounted at the center of the test board. One port of the vector network analyzer (VNA) is connected to the microstrip line, and the other port is connected to the loop probe. The other end of the microstrip line is terminated by 50 Ω. The NSE is defined by a ratio of two-port S-parameters between the microstrip line and loop probe as $\text{NSE} = 20\,log_{10}\left(S_{21}^{R}/S_{21}^{L}\right)$, where S_{21}^{R} and S_{21}^{L} are transmission coefficients of S-parameters in the absence (reference) and presence (load) of the magnetic sheet, respectively. The measured permeabilities and NSE were used as input data for Minitab's regression analysis to obtain the regression models.

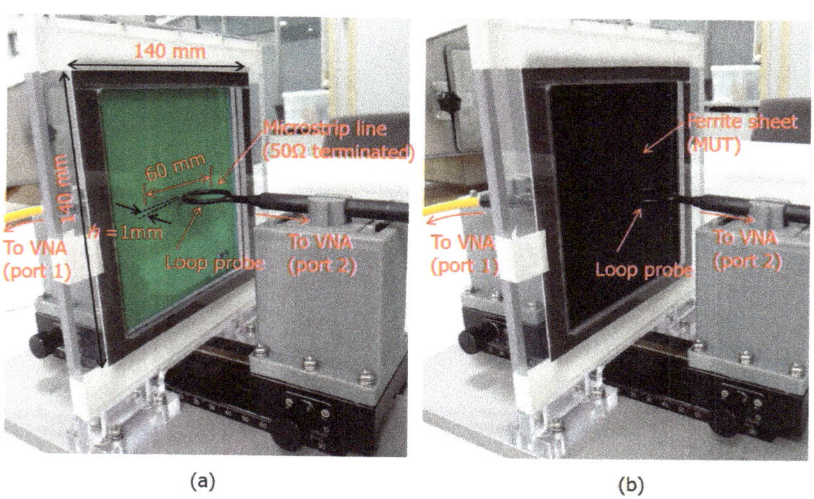

Figure 4. Photos of fabricated test fixture for magnetic shielding measurement. (**a**) Reference measurement. (**b**) Load measurement.

Figure 5 shows the complex relative permeability and near-field magnetic SE (NSE) of the R2 ferrite sheet with 0.4 mm thickness measured up to 100 MHz. The real part of the relative permeability remains nearly constant up to 10 MHz, peaks at 20 MHz, and then decreases linearly at higher frequencies. The imaginary part of the permeability is close to zero up to 10 MHz, increases up to approximately 30 MHz, and then decreases. The measured NSE closely follows the changes in the real permeability of the MUT as a function of frequency. This indicates that the NSE is determined by the value of the real permeability and can be predicted using the real permeability value.

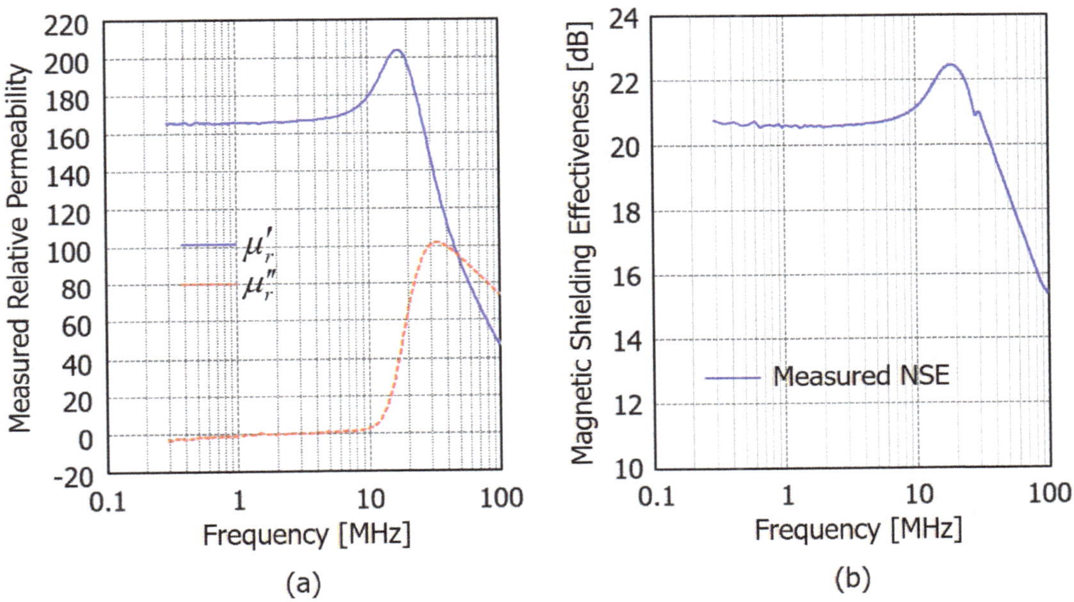

Figure 5. Measurement results of R2. (**a**) Relative permeability; (**b**) Near-field magnetic shielding effectiveness (NSE).

Eight regression models were derived using the measured real relative permeability and NSE of five ferrite sheets in Group #1. Model #1 is a linear combination of frequency and real permeability. Model #2 adds a square term of the permeability to this linear combination and Model #3 includes up to a cubic term of the permeability. Model #4 includes the fourth-order term of permeability. Model #5 incorporates a square term of the frequency into the linear combination of frequency and real permeability, and Model #6 includes both the square terms of frequency and permeability. Model #7 includes the third-order term of permeability from Model #6, while Model #8 includes the fourth-order term of permeability from Model #7. All models contain a linear term of thickness. As a result, these models demonstrate the impact of higher-order terms of permeability and frequency on the prediction accuracy of NSE. These models are determined by the coefficients in front of each term, which are derived through Minitab's regression analysis, and the results are listed in Table 2.

Table 2. Extracted eight regression models, their adjusted coefficient of determination (R-Sq(adj)), and p-values in regression analysis.

Models	Prediction Equations of NSE *	R-Sq(adj)	p-Value
#1	$NSE = 9.03 - 0.0345 \times f^* + 23.2 \times t^* + 0.0209 \times \mu_r$	89.2%	0.00
#2	$NSE = 4.31 + 0.00068 \times f + 25.2 \times t + 0.0451 \times \mu_r - 0.285 \times 10^{-4} \times (\mu_r)^2$	92.6%	0.00
#3	$NSE = 0.70 + 0.0234 \times f + 25.5 \times t + 0.08 \times \mu_r - 0.134 \times 10^{-3} \times (\mu_r)^2 - 0.902 \times 10^{-7} \times (\mu_r)^3$	93.3%	0.00
#4	$NSE = -3.14 + 0.042 \times f + 25.5 \times t + 0.135 \times \mu_r - 0.411 \times 10^{-3} \times (\mu_r)^2 + 0.641 \times 10^{-6} \times (\mu_r)^3 - 0.364 \times 10^{-9} \times (\mu_r)^4$	93.9%	0.00
#5	$NSE = 7.49 + 0.05237 \times f - 0.0941 \times 10^{-2} \times f^2 + 24.3 \times t + 0.0227 \times \mu_r$	91.7%	0.00
#6	$NSE = 2.74 + 0.0881 \times f - 0.0946 \times 10^{-2} \times f^2 + 26.3 \times t + 0.047 \times \mu_r - 0.286 \times 10^{-4} \times (\mu_r)^2$	95.2%	0.00
#7	$NSE = 1.05 + 0.0926 \times f - 0.0872 \times 10^{-2} \times f^2 + 26.4 \times t + 0.0644 \times \mu_r - 0.818 \times 10^{-4} \times (\mu_r)^2 + 0.453 \times 10^{-7} \times (\mu_r)^3$	95.4%	0.00
#8	$NSE = 0.42 + 0.0926 \times f - 0.0834 \times 10^{-2} \times f^2 + 26.4 \times t + 0.0739 \times \mu_r - 0.129 \times 10^{-3} \times (\mu_r)^2 + 0.136 \times 10^{-6} \times (\mu_r)^3 - 0.58 \times 10^{-10} \times (\mu_r)^4$	95.4%	0.00

* NSE is in dB, f is in MHz, and t is in mm.

When performing regression analysis, we should check two output results as criteria that determine the reliability of the analysis: p-value and adjusted coefficient of determination (R-Sq(adj)). The p-value measures the probability of obtaining the observed results, assuming that the null hypothesis is true. The lower the p-value, the greater the statistical significance of the observed difference. If insignificant independent variables are included in the regression equation, the R-Sq(adj) decreases and is often used as a criterion for selecting the optimal model. If the p-value is smaller than 0.05 and R-Sq(adj) is large, it means that the regression equation is statistically significant and accurate. Table 2 also lists the R-Sq(adj) and p-value for each model. It can be seen that the p-value for all models is 0.00, and the R-Sq(adj) for all models, except for Model #1, is above 90%, with R-Sq(adj) increasing to 95.4% in Models #7 and #8. This indicates that the extracted regression models are statistically significant and accurate. As higher-order terms of the permeability are included, the R-Sq(adj) increases, but models with the fourth-order term in permeability show similar R-Sq(adj) values to those with the third-order term. Therefore, it means that the fourth-order term of permeability is not necessary. Additionally, models that include the squared frequency term have a higher R-Sq(adj) compared to those that do not.

Figure 6 shows the regression analysis results for Model #6 and the residual plots for the predicted NSE. A normal probability plot of the residuals is used to check whether the residuals follow a normal distribution. The residuals histogram helps identify whether the data are skewed in a specific direction or if there are any outliers. The residuals versus fits plot is used to verify if the residuals are randomly distributed. If the plot shows a random distribution without any consistent pattern, it suggests that the assumptions of linearity and homoscedasticity are met. The residuals versus order plot checks whether the residuals are independent over time; if a specific pattern appears over time, it may indicate the presence of autocorrelation, suggesting that time-dependent factors are not accounted for in the regression model. Consequently, the residuals for the NSE in Model #6 are randomly distributed, follow a normal distribution, have no outliers, and are independent over time.

Regression Analysis: NSE [dB] versus Freq[MHz], Freq2, TH [mm], MU, MU2

The regression equation is

$$NSE\ [dB] = 2.74 + 0.0881\ Freq[MHz] - 0.0946\times10^{-2}\ Freq2 + 26.3\ TH\ [mm] + 0.0470\ MU - 0.00286\times10^{-2}\ MU2$$

Predictor	Coef	SE Coef	T	P
Constant	2.7424	0.6203	4.42	0.000
Freq[MHz]	0.08811	0.01092	8.07	0.000
F2	−0.09461	0.01057	−8.95	0.000
TH [mm]	26.3421	0.8135	32.38	0.000
MU	0.047011	0.002435	19.31	0.000
MU2	−0.0028645	0.0002766	−10.36	0.000

S = 0.817365 R-Sq = 95.4% R-Sq(adj) = 95.2%

Analysis of Variance

Source	DF	SS	MS	F	P
Regression	5	1988.14	397.63	595.18	0.000
Residual Error	144	96.20	0.67		
Total	149	2084.35			

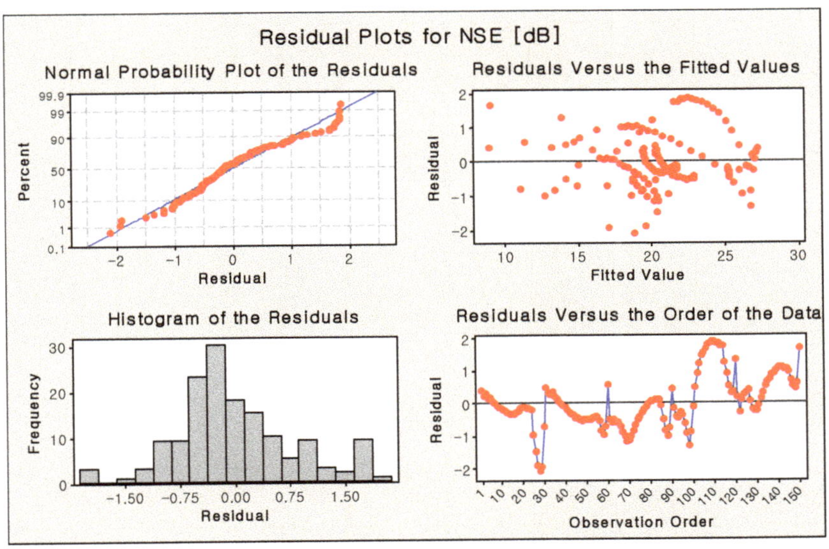

Figure 6. Regression analysis results for Model #6 and the residual plots for the predicted NSE.

3. Comparison and Error Analysis

3.1. Comparison

The predicted NSE for the five ferrite sheets in Group #1 and the two ferrite sheets in Group #2 listed in Table 1 were calculated using the eight regression models. Figure 7 compares the predicted and measured NSE as a function of frequency for R1 and R2 in Group #1, and P1 and P2 in Group #2. As previously mentioned, the NSE characteristics up to 100 MHz can be broadly divided into three regions: a region where NSE remains constant below 10 MHz, a region around 20 MHz with a hill-like shape, and a region above 20 MHz where NSE decreases linearly. In the constant NSE region, all prediction models, except for Model #1, appear to show an error within 1 dB compared to the measured NSE. In the hill region around 20 MHz, prediction models other than Model #1 provide predictions that closely match the measured values. In the linearly decreasing region above

20 MHz, models with only the first-order frequency term (Model #1, #2, #3, and #4) appear to be more accurate than those including the second-order frequency term (Model #5, #6, #7, and #8). Anyway, the regression models provide predicted NSE that reasonably match the measured results for both the MUTs used in the model extraction and those not used in the extraction process. Consequently, various commercial magnetic sheets are verified and consistency is shown at the engineering level within 2 dB difference. It is confirmed that the proposed regression models can accurately predict the NSE of ferrite sheets using only the measured permeability values up to 100 MHz.

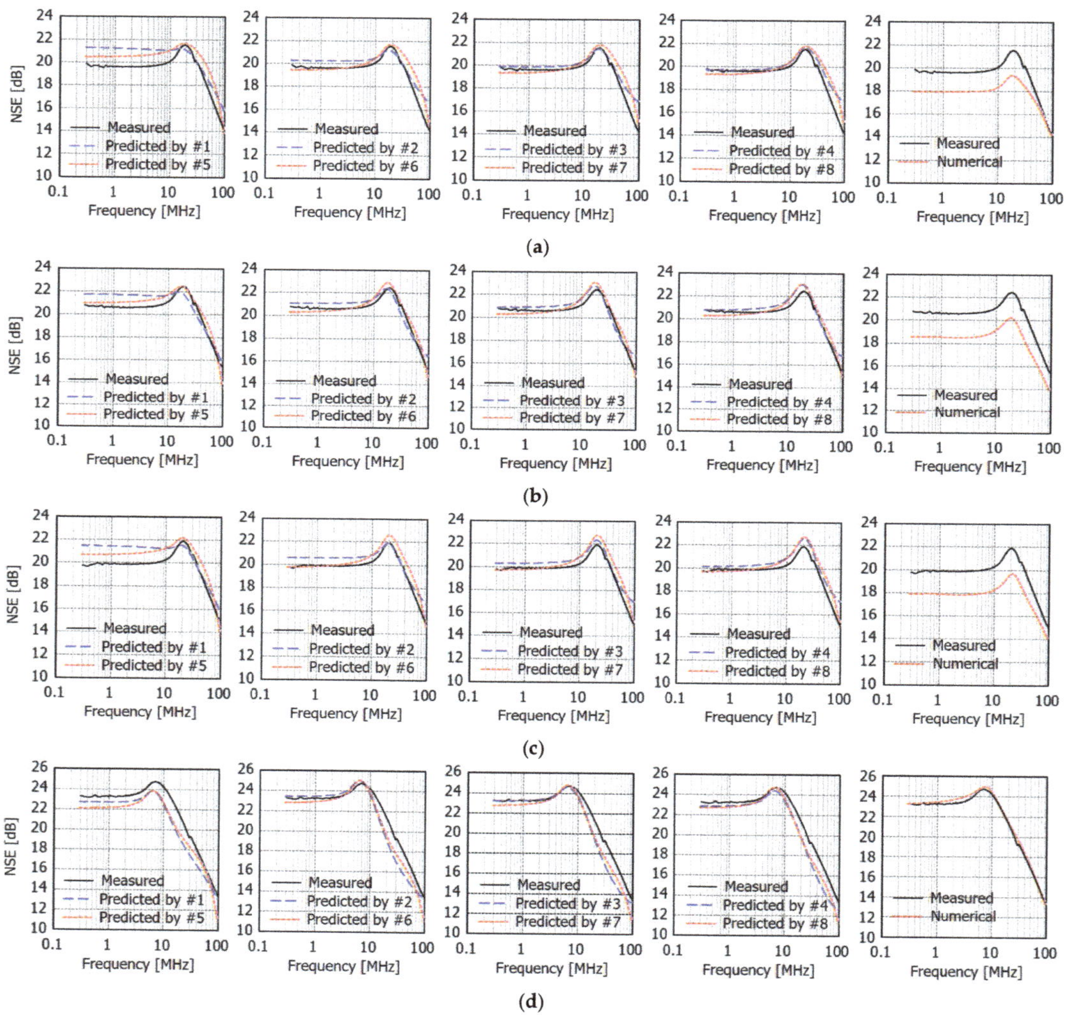

Figure 7. Comparison of predicted and measured NSE versus frequency. **(a)** R1; **(b)** R2; **(c)** P1; **(d)** P2.

Additionally, the measured permeability was imported into CST Microwave Studio [14] to perform numerical simulations, and the results were compared with the measured NSE as shown in Figure 7. For R1, R2, and P1, the numerical simulation results were approximately 2 dB lower than the measured NSE values. In the case of P2, the numerical simulation results closely matched the measured values. Although it was challenging to pinpoint the exact cause of the discrepancies between the numerical and measured values, it was likely due to differences between the actual permeability of the magnetic sheets and the measured permeability, as well as discrepancies between the numerical simulation setup and the measurement conditions.

3.2. Error Analysis

To quantitatively analyze the difference or error between the predicted NSE and the measured NSE, the mean square error (MSE) was calculated [15] and compared. Figure 8 compares the accuracy of the eight regression models by calculating the MSE between the predicted and measured NSE for the five MUTs in Group #1 (for regression) and the two MUTs in Group #2 (for prediction) separately. As shown in Table 2, Models #1, #2, #3, and #4 include only the linear term for frequency, while Models #5, #6, #7, and #8 include the quadratic term for frequency as well. From Figure 8, it can be observed that models incorporating the quadratic term for frequency exhibit smaller MSE values. For models with only the linear term for frequency, the inclusion of higher-order permeability terms reduces the MSE. However, higher-order terms beyond the quadratic term do not significantly decrease the MSE. This trend also applies to models that include the quadratic term for frequency. In conclusion, the simplest predictive model with the smallest MSE is Model #6. For models with only the linear frequency term, the average MSE for Group #1 (Regression) and Group #2 (Prediction) fluctuates depending on the inclusion of higher-order permeability terms. However, for models incorporating the quadratic frequency term, the average MSE for Group #1 (Regression) is consistently smaller than that for Group #2 (Prediction). Model #6 achieves the smallest total average MSE across all MUTs.

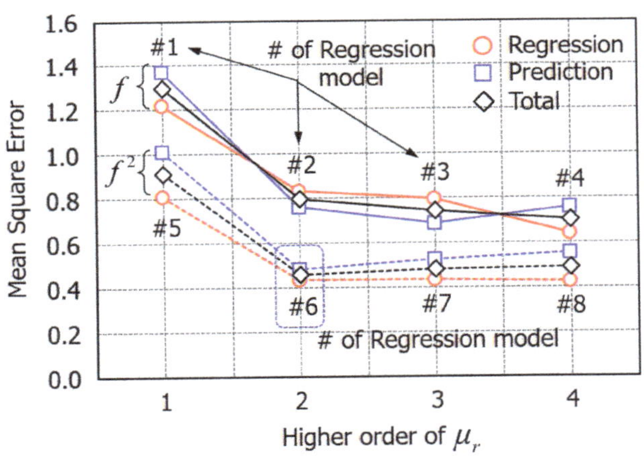

Figure 8. Mean square errors (MSEs) of MUTs according to the eight regression models.

Table 3 numerically presents the values shown in Figure 8. The values with the smallest MSE among the 8 models in each category (Regression, Prediction, and Total) have been bolded. Among the average MSEs of the MUTs in Group #1, which were used to extract the regression models, the smallest value is 0.426 for Model #8. For the MUTs in Group #2, which were used exclusively for prediction, the smallest average MSE is 0.484 for Model #6. The total average MSE for all models, except Model #1, is less than 1, with Model #6 having the smallest MSE of 0.46. In the case of numerical simulation, the average MSE for the MUTs in Group #1 is significantly higher, at 9.659, which is also evident in Figure 7. In contrast, the average MSE for the MUTs in Group #2 is 1.236. The total average MSE for numerical simulation is 5.446, which is ten times larger than that of Model #6.

In conclusion, the prediction accuracy of the regression models is far superior to that of numerical simulation.

Table 3. Comparison of averaged MSEs of extracted eight regression models and numerical simulation.

Models	Average Mean Square Errors (MSEs)		
	Regression (Group #1)	Prediction (Group #2)	Total
#1	1.218	1.370	1.294
#2	0.833	0.761	0.797
#3	0.797	0.689	0.743
#4	0.643	0.760	0.702
#5	0.810	1.013	0.912
#6	0.435	**0.484**	**0.460**
#7	0.436	0.526	0.481
#8	**0.426**	0.556	0.491
Numerical	9.656	1.236	5.446

Lastly, the time required to calculate NSE using regression models was compared with the time required using numerical simulations. Here, the time to extract regression models was excluded; only the calculation time for NSE, given permeability data, was considered. For regression models, the calculations involve only basic arithmetic operations such as addition, subtraction, and multiplication, allowing results to be obtained almost instantaneously. In contrast, numerical simulations require extensive meshing due to the high permeability of the magnetic sheets. As a result, using a standard PC (Intel Core i7-6800K, 64 GB) without an accelerator, the simulation takes several hours to complete. Table 4 presents the NSE computation time for the smallest case among the seven sheets listed in Table 1. In conclusion, from a computation time perspective, regression models are significantly more efficient for NSE prediction compared to numerical simulation methods.

Table 4. Comparison of computation time.

Methods	Computation Time
Regression Models	<1 s
Numerical Simulation	>4 h 31 m 9 s

4. Conclusions

In this paper, we presented a method to predict the near-field magnetic shielding effectiveness (NSE) of ferrite sheets using the prediction models obtained through regression analysis. The commercial ferrite sheets used in the analysis were divided into Group #1 and Group #2. Eight regression models were developed using the measured relative permeability and NSE of the sheets in Group #1. The predicted NSE of the sheets in both Group #1 and Group #2 were then calculated and compared with the measured NSE. To verify the accuracy of the proposed regression models, an analysis of mean square errors (MSEs) was employed. The regression models with second-order frequency terms exhibited lower MSE compared to those with only first-order frequency terms. Additionally, while the inclusion of higher-order permeability terms further reduced MSE, the improvement was not significant beyond quadratic terms. Among the eight regression models, Model #6, which includes quadratic terms for both frequency and permeability, achieved the smallest MSE of 0.46. This is less than one-tenth of the MSE from numerical simulations. In terms of computation time, numerical simulations require over 4 h, whereas the proposed regression models can predict NSE in real time. Therefore, the regression models provide a highly effective method for NSE prediction compared to numerical simulations. As a result, if the real permeability and thickness information of ferrite sheets are known, the NSE can be predicted using regression models without the need for NSE measurements.

Finally, we would like to point out the limitations of the NSE prediction method proposed in this paper using regression analysis. First, since regression analysis is a statistical technique, the accuracy of the predictive model improves as the amount of data increases. Therefore, it is crucial to secure a sufficient number of MUTs. Second, the accuracy of the data used to extract the regression models is essential. If the data contains outliers, the prediction accuracy of the extracted model will decrease. Thus, for accurate NSE prediction, the precise measurement of the permeability of magnetic sheets is paramount. Not only the accurate measurement of the NSE of magnetic sheets but also the reproducibility and repeatability of the measurement methods for precise permeability measurements must be secured first.

Author Contributions: Conceptualization, H.H.P.; methodology, H.H.P.; validation, H.H.P.; formal analysis, H.H.P.; investigation, H.H.P. and H.L.; resources, H.L. and D.-K.H.; data curation, H.H.P. and H.L.; writing—original draft preparation, H.H.P.; writing—review and editing, H.H.P. and H.L.; supervision, H.H.P. and D.-K.H.; project administration, D.-K.H.; funding acquisition, H.L. and D.-K.H. All authors have read and agreed to the published version of the manuscript.

Funding: This work was supported by the Hyundai MOBIS Co., Ltd. (No. CTR220600648).

Data Availability Statement: Data are contained within the article.

Conflicts of Interest: Authors H.L. and D.-K.H. were employed by the company Hyundai MOBIS Co., Ltd. The remaining author declares that the research was conducted in the absence of any commercial or financial relationships that could be construed as a potential conflict of interest.

References

1. Yoshida, S.; Ono, H.; Ando, S.; Tsuda, F.; Ito, T.; Shimada, Y.; Yamaguchi, M.; Arai, K.; Ohnuma, S.; Masumoto, T. High-frequency noise suppression in downsized circuits using magnetic granular films. *IEEE Trans. Magn.* **2001**, *37*, 2401–2403. [CrossRef]
2. Watanabe, K.; Jike, K.; Tanaka, S.; Miura, N.; Nagata, M.; Takahashi, A.; Miyazawa, Y.; Yamaguchi, M. Magnetic composite sheets in IC chip packaging for suppression of undesired noise emission to wireless communication channels. In Proceedings of the 2019 12th International Workshop on the Electromagnetic Compatibility of Integrated Circuits (EMC Compo), Hangzhou, China, 21–23 October 2019; pp. 219–221.
3. Deutschmann, B.; Khan, J.S.; Winkler, G.; Victoria, J. Reduction of electromagnetic emission from ICs using soft flexible ferrite sheets. In Proceedings of the 2021 13th International Workshop on the Electromagnetic Compatibility of Integrated Circuits (EMC Compo), Bruges, Belgium, 9–11 March 2022; pp. 19–24.
4. Park, H.H. Reduction of electromagnetic noise coupling to antennas in metal-framed smartphones using ferrite sheets and multi-via EBG structures. *IEEE Trans. Electromagn. Compat.* **2018**, *60*, 394–401. [CrossRef]
5. DeRoy, P.; Koledintseva, M.Y.; Rostamzadeh, C.F.; Worden, M.; Monti, T.; Schulz, W.; Ramanujan, A. Effectiveness of noise suppressing sheet material for mitigation of automotive radiated emissions. *IEEE Trans. Electromagn. Compat.* **2021**, *63*, 398–409. [CrossRef]
6. Zhang, M. Low-frequency magnetic fields in electric vehicles: Challenges, shielding, and design considerations. Compliance Magazine 31 March 2022. Available online: https://incompliancemag.com/low-frequency-magnetic-fields-in-electric-vehicles/ (accessed on 12 August 2024).
7. Jobava, R.G.; Gheonjian, A.L.; Hippeli, J.; Chiqovani, G.; Karkashadze, D.D.; Bogdanov, F.G.; Khvitia, B.; Bzhalava, A.G. Simulation of low-frequency magnetic fields in automotive EMC problems. *IEEE Trans. Electromagn. Compat.* **2014**, *56*, 1420–1430. [CrossRef]
8. MINITAB Release 14.20.0. Available online: https://www.minitab.com/en-us/ (accessed on 14 October 2024).
9. Keysight Technologies, Keysight 16454A Magnetic Material Test Fixture, Operation and Service Manual. 6 September 2017. Available online: https://www.keysight.com/us/en/assets/9018-01409/user-manuals/9018-01409.pdf (accessed on 12 August 2024).

10. Chen, C.-H.; Tseng, Y.-C.; Lin, I.-C.; Fu, C.-C.; Liao, K.-H.; Wu, T.-L. Transmission-line based modeling for conformal shielding in advance system-in-package (SiP). In Proceedings of the 2015 Asia-Pacific Symposium on Electromagnetic Compatibility (APEMC), Taipei, Taiwan, 26–29 May 2015; pp. 521–523.
11. Sano, E.; Ikebe, M. Analytical methods for evaluating the characteristics of noise suppression sheets. *IEEE Trans. Electromagn. Compat.* **2022**, *64*, 1052–1057. [CrossRef]
12. Park, H.H.; Kim, C. Wideband magnetic field shielding measurement method of EMI sheets using microstrip lines. *J. Korean Inst. Electromagn. Eng. Sci.* **2023**, *34*, 554–560. [CrossRef]
13. *RF-R 400-1*; H-Field Probe 30 MHz up to 3 GHz. Langer EMV-Technik GmbH: Bannewitz, Germany.
14. Microwave Studio (MWS). CST. 2020. Available online: https://www.3ds.com/products/simulia/cst-studio-suite (accessed on 15 August 2024).
15. CPS Tech Inc. Available online: http://x-mate.cloud/sources/charts.html (accessed on 24 December 2024).

Disclaimer/Publisher's Note: The statements, opinions and data contained in all publications are solely those of the individual author(s) and contributor(s) and not of MDPI and/or the editor(s). MDPI and/or the editor(s) disclaim responsibility for any injury to people or property resulting from any ideas, methods, instructions or products referred to in the content.

electronics

MDPI

Article

Investigation of Electrical Performance Degradation of β-Ga_2O_3 Schottky Barrier Diodes Induced by X-Ray and Neutron Irradiation

Pengfei Zhao [1], Xu Tan [2], Weili Fu [2,*] and Teng Ma [2,*]

1 School of Microelectronics, Faculty of Electronic and Information Engineering, Xi'an Jiaotong University, Xi'an 710049, China
2 Science and Technology on Reliability Physics and Application of Electronic Component Laboratory, China Electronic Product Reliability and Environmental Testing Research Institute, Guangzhou 511370, China
* Correspondence: kobefwl@163.com (W.F.); mateng_ceprei@163.com (T.M.)

Abstract: This paper presents a preliminary investigation into the total dose effects and displacement damage effects on β-Ga_2O_3 Schottky barrier diodes (SBDs) induced by X-rays with an average energy of 8–20 keV and 1 MeV reactor neutrons. The electrical performance of the devices before and after irradiation was evaluated through direct current (I-V) and capacitance–voltage (C-V) measurements. The results indicate that under X-ray irradiation, as the irradiation fluence increases, the forward current density, leakage current, and reverse current density of the devices increase, suggesting a progressive degradation of device performance with higher irradiation fluence. In the case of neutron irradiation, the forward current density decreases, while the leakage current and reverse current density increase with rising irradiation fluence. By employing techniques such as low-frequency noise (LFN) and deep-level transient spectroscopy (DLTS), changes in defect concentrations before and after irradiation were analyzed. It was found that the primary causes of device performance degradation are the interface defects induced by X-ray irradiation and the increased bulk defect concentration caused by neutron irradiation. These findings were further validated through two-dimensional numerical simulations using TCAD tools, providing significant theoretical insights and experimental data to enhance reliability and optimize the design of such devices.

Keywords: β-Ga_2O_3 SBD; X-ray irradiation; neutron irradiation; defects

1. Introduction

β-Ga_2O_3 SBDs for high-power space applications: Radiation effects and degradation mechanisms of β-Ga_2O_3, with its ultra-wide bandgap and high breakdown electric field, exhibit exceptional properties that enable it to withstand high voltages, making it highly suitable for high-power electronic devices in spacecraft, such as power amplifiers and power management modules. These devices can significantly enhance the energy utilization efficiency of spacecraft. The Baliga's figure of merit (BFOM) for β-Ga_2O_3 is as high as 3214, approximately ten times that of SiC and four times that of GaN. This indicates that devices fabricated using β-Ga_2O_3 will exhibit lower conduction losses and higher power conversion efficiency, making them ideal for high-voltage and high-power environments in spacecraft [1–5].

However, in space applications, the reliability of β-Ga_2O_3 SBDs is critically affected by radiation effects caused by the harsh space radiation environment. The impact of various high-energy rays and particles in space cannot be overlooked. Due to its high density, thermal stability, chemical stability, and strong theoretical breakdown electric field, β-Ga_2O_3 possesses unique advantages in X-ray detection [6–9]. Therefore, investigating the radiation effects of X-rays on β-Ga_2O_3 SBDs is essential.

Neutrons, which are electrically neutral and widely distributed in space, primarily cause displacement damage in β-Ga_2O_3 SBDs. Previous studies on displacement damage in β-Ga_2O_3 have mainly focused on the material itself. For instance, Polyakov et al. [10]. observed an increase in deep trap states at $E_C-0.74$ eV, $E_C-1.05$ eV, and $E_C-1.20$ eV in silicon-doped β-Ga_2O_3 after 1.25 MeV neutron irradiation. Similarly, Cojocaru et al. [11]. found that fast neutron irradiation leads to the formation of gallium vacancies, thereby reducing the conductivity of β-Ga_2O_3. These findings raise questions about whether β-Ga_2O_3 SBDs can fully leverage the radiation resistance of β-Ga_2O_3 material. Thus, investigating the degradation mechanisms of β-Ga_2O_3 SBDs under neutron irradiation is crucial for evaluating their resistance to displacement damage [12–14].

In this work, the total dose effects of β-Ga_2O_3 SBDs under X-ray irradiation at different doses and the displacement damage effects under neutron irradiation at various fluences were systematically studied. The degradation mechanisms induced by X-ray and neutron irradiation were analyzed using low-frequency noise (LFN) and deep-level transient spectroscopy (C-DLTS) techniques. The conclusions were further validated through TCAD simulation tools, providing a comprehensive understanding of the radiation-induced degradation in β-Ga_2O_3 SBDs and offering valuable insights for improving their reliability in space applications.

2. Materials and Methods

Figure 1 illustrates the cross-sectional schematic of the device. The epitaxial layer of the device was grown using halide vapor phase epitaxy (HVPE) technology, with a thickness of 10 µm and a doping concentration of 3.0×10^{16} cm^{-3}. The substrate of the device consists of a 650 µm thick β-Ga_2O_3 single crystal. The cathode employs a Ti/Au (20/200 nm) ohmic contact, deposited via electron beam evaporation and subjected to rapid thermal annealing in a nitrogen (N_2) atmosphere to significantly enhance the ohmic contact properties. The anode, on the other hand, utilizes a Ni/Au (50/200 nm) Schottky contact fabricated through a lift-off process. The entire device is packaged in a dual-pin TO configuration with an anode area of 4 mm^2.

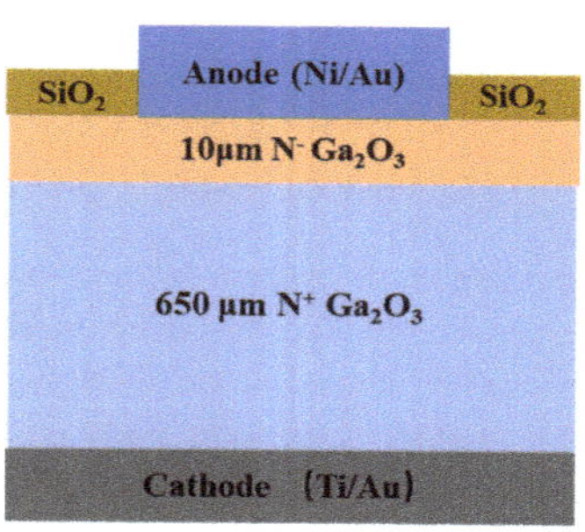

Figure 1. Cross-sectional schematic of the β-Ga_2O_3 SBD device structure.

The high-temperature thermal oxidation process effectively reduces the oxygen vacancy concentration at the edge of the Schottky contact; the thermal oxidation treatment was conducted at a temperature of 400 °C for a duration of 20 min, thereby significantly lowering the peak electric field at the device edges and substantially increasing the breakdown voltage. The breakdown voltage of this SBD is approximately 650 V. This optimized design not only improves the device's electrical performance but also enhances its reliability for high-power applications.

The X-ray irradiation experiments were conducted at the Xinjiang Technical Institute of Physics and Chemistry, Chinese Academy of Sciences. The experiments were performed at room temperature, with the device pins grounded. The X-rays used had an average energy ranging from 8 to 20 keV, a dose rate of 50 rad(Si)/s, and cumulative doses of 500 krad(Si), 1 Mrad(Si), and 2 Mrad(Si). There are a total of ten samples for the X-ray irradiation experiment, with one serving as a control and three each for the other three different irradiation fluences used in the experiment, with the experimental results being the average values. Noise information was extracted using the LS-PRO system before and after irradiation.

The neutron irradiation experiments were carried out at the Northwest Institute of Nuclear Technology. These experiments were also conducted at room temperature, with the device pins grounded. The devices were irradiated with 1 MeV neutrons at a fluence of 1×10^{13} n/cm^2, the neutron irradiation fluence rate is 1×10^9 n/cm^2, and the irradiation time is 1000 s. After irradiation, bulk defect information was obtained using DLTS.

These experiments provide critical insights into the radiation-induced degradation mechanisms of β-Ga_2O_3 SBDs under both X-ray and neutron irradiation, contributing to a better understanding of their performance and reliability in radiation-rich environments.

3. Results and Discussion

A. X-ray Irradiation

Figure 2a,b illustrate the forward and reverse electrical characteristics of the device before and after X-ray irradiation; in the figure, "Fresh" represents the sample before irradiation, and the three different doses are 500 krad, 1 Mrad, and 2 Mrad. It is evident from the figures that both the forward and reverse current densities of the device increase with the increasing irradiation dose. The forward current of the device is primarily associated with the thermionic emission mechanism related to the Schottky barrier height. Based on this mechanism, the ideal factor and Schottky barrier height of the device after X-ray irradiation were calculated using Equation (1) and Equation (2), respectively. In these equations, J represents the current density, A is the anode area of the device, T is the thermodynamic temperature (300 K at room temperature), $q = 1.6 \times 10^{-19}$ C is the electron charge, k is the Boltzmann constant, n is the ideality factor of the SBD, J_s is the reverse saturation current density, A^* is the Richardson constant, which is typically 40.8 A/(cm^2·K^2) for β-Ga_2O_3 material, and ϕ is the Schottky barrier height [14–16].

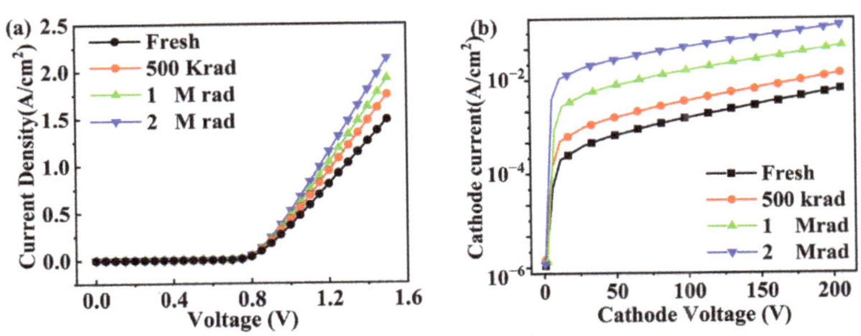

Figure 2. Variations in the electrical properties of the device after X-ray irradiation: (**a**) forward current density; (**b**) reverse current density.

$$J = A^{*}T^{2}\exp\left(-\frac{q\phi}{kT}\right)\exp\left(-\frac{qV}{nkT}\right) = J_{s}exp\left(-\frac{qV}{nkT}\right) \quad (1)$$

$$\Phi = \frac{kT}{q}\ln\left(\frac{AA^{*}T^{2}}{J_{S}}\right),\ n = \frac{q}{kT}\frac{\partial V}{\partial(lnJ)} \quad (2)$$

The calculations indicate that the Schottky barrier height ϕ decreases with the increasing irradiation dose, while the ideality factor n decreases with the increasing irradiation dose. The results are shown in Table 1. The reduction in the Schottky barrier height leads to an increase in the forward current density of the device, and simultaneously, it also increases the reverse current density.

Table 1. Changes in device electrical parameters before and after irradiation.

n (Before)	*n* (After)	*n*	*Φ* (Before)	*Φ* (After)	*Φ*
1.126	1.108	−1.59%	1.09	1.04	−4.58%
1.170	1.137	−2.82%	1.17	1.08	−7.69%
1.212	1.183	−2.39%	1.13	1.06	−6.19%
1.188	1.162	−2.19%	1.11	1.07	−5.41%

The reverse current of the SBD is not only related to the current due to thermionic emission but also closely associated with the change in the carrier concentration in the epitaxial layer due to the variation in the concentration of internal defects in the material.

C-V measurement is a technique used to characterize the electrical properties of semiconductor materials or devices by analyzing the relationship between capacitance and voltage. It is particularly effective for determining the carrier concentration in SBDs. Under reverse bias conditions, the capacitance of an SBD is closely related to the carrier concentration. As the reverse bias voltage changes, the width of the depletion region varies accordingly, and this variation is reflected in the capacitance. By measuring these capacitance changes, the carrier concentration can be accurately inferred. Consequently, to evaluate the impact of irradiation on the device, we utilize the C-V characteristic curves of the β-Ga_2O_3 SBD before and after irradiation to calculate the changes in carrier concentration within the epitaxial layer. The C-V variation is shown in Figure 3. The calculation process follows Equations (3) and (4) [17–19]. In these equations, N_D represents the carrier concentration, and ϵs is the relative permittivity of β-Ga_2O_3, which is $10.2\varepsilon_0$, where ε_0 is the vacuum permittivity with a value of 8.85×10^{-12} F/m.

$$\frac{1}{C^2} = \frac{2}{q\varepsilon_s N_D A^2}(V_{bi} - V) \quad (3)$$

$$N_D = -\frac{2}{q\varepsilon_s A^2}\left(\frac{d(1/C^2)}{dV}\right)^{-1} \quad (4)$$

The calculated results of the carrier concentration variation in the epitaxial layer of the device are shown in Table 2.

Table 2 shows that the carrier concentration of the device decreases significantly as the irradiation dose increases. When X-rays strike the device, they generate numerous electron–hole pairs in its epitaxial layer. Since electrons have higher mobility than holes, the generated holes tend to remain in the epitaxial layer and accumulate at the interface between the metal and semiconductor, forming interface defects. These defects capture carriers, leading to a reduction in the carrier concentration of the device.

Table 2. Carrier concentration before and after irradiation in the drift layer.

	N_d (Before Irradiation)	N_d (After Irradiation)	Change
Fresh	8.410×10^{15}	8.196×10^{15}	−2.54%
500 Krad	8.206×10^{15}	7.910×10^{15}	−3.61%
1 Mrad	8.124×10^{15}	7.707×10^{15}	−5.13%
2 Mrad	8.138×10^{15}	7.575×10^{15}	−6.92%

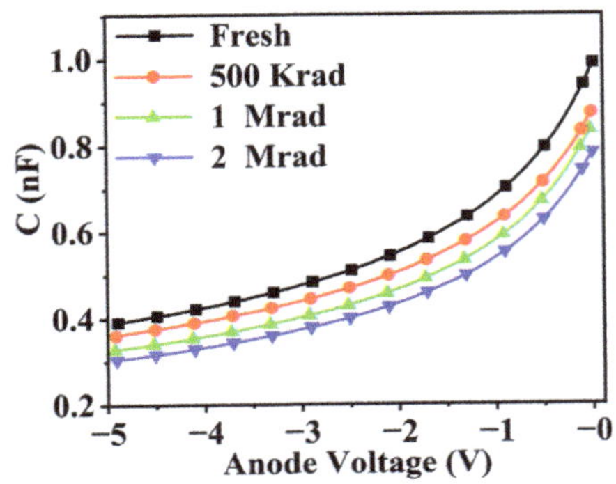

Figure 3. Variation of C-V curves of the device under different X-ray irradiation doses.

Electrical characteristics provide insight into the internal structure and current conduction mechanisms of electronic devices, with changes in these characteristics indicating structural alterations. X-ray irradiation causes TID effects that damage β-Ga_2O_3 SBDs, leading to changes in their electrical properties, such as a decrease in the ideality factor and an increase in forward current density after the device is activated. Although these electrical parameters can characterize the radiation resistance of β-Ga_2O_3 diodes, their minimal changes make them less sensitive indicators of irradiation-induced alterations.

In contrast, noise is increasingly recognized as a highly sensitive method for detecting potential internal defects in devices. It offers a rapid and non-destructive testing approach [20,21]. When changes occur in the device's original lattice structure, electron state density, or impurity distribution over time or under stress, noise detection exhibits more significant variations compared to other structure-sensitive parameters [22]. Noise in electronic devices is highly sensitive to these defects, making noise testing a common method for diagnosing and predicting defect changes [23]. Numerous studies have shown that $1/f$ noise serves as a sensitive parameter reflecting defect information in electronic components [24]. Therefore, when studying defects and interface state changes in the space charge region and near the oxide–semiconductor interface after irradiation, noise diagnostic techniques more effectively reflect the extent of radiation damage than traditional electrical parameters, as shown in Figure 4.

The current noise density spectrum induced by interface defects can be expressed using the formula proposed by Jantsch et al. [25]. The expression for the current noise power spectral density is as follows.

$$S_{I_r} = \frac{G\beta}{f}\left(\frac{qI}{4\varepsilon_S}\right)^2 \frac{qD_{is}}{\pi N_d WA} \tag{5}$$

The current noise power spectral density can be described by the following equation proposed by Jantsch et al. [25], where G is a constant with a value of 0.1, D_{is} is the density of interface states, $q = 1.6 \times 10^{-19}$ C is the electron charge, k is the Boltzmann constant, and β is a parameter describing the carrier transport mechanism in the Schottky barrier diode

(SBD). For the thermionic emission mechanism, $\beta = qkT$, f is the frequency at which the noise is measured, I is the forward current during the measurement, ε_S is the permittivity of β-Ga_2O_3, N_d is the doping concentration after irradiation, h is the Planck constant, W is the width of the space-charge region, and A is the effective contact area of the device.

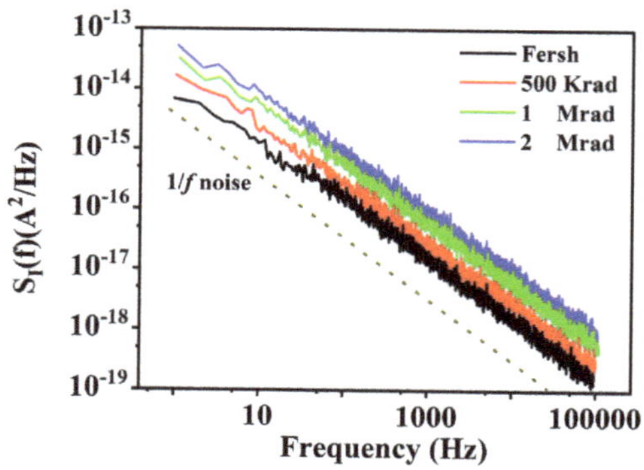

Figure 4. Variation of low-frequency noise power spectral density of the device with irradiation dose.

It is evident that besides the doping concentration, the density of interface states is a key factor affecting noise power spectral density. At higher irradiation doses, the concentration of interface defects within the device increases, leading to a rise in noise power spectral density.

B. Neutron Irradiation

Figure 5 illustrates the changes in the device's forward and reverse electrical performance before and after neutron irradiation; in the figure, "Fresh" represents the sample before irradiation, and the neutron irradiation fluence is 1×10^{13} n/cm^2, denoted as 1×10^{13} in the figure.

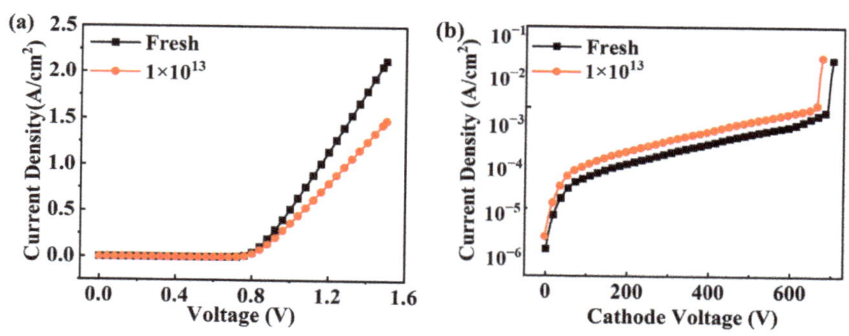

Figure 5. Variations in the electrical properties of the device before and after neutron irradiation: (**a**) forward current density; (**b**) reverse current density.

As shown in Figure 5, the electrical characteristics of the device after neutron irradiation differ from those after X-ray irradiation. The forward current density of the device decreases while the reverse current density increases. Calculations using Equations (1) and (2) reveal that the SBH of the device decreases from 1.12 eV to 1.09 eV, and the ideality factor decreases from 1.17 to 1.10, with relatively small changes. In the research by Yue et al. [26]., it was suggested that the degradation of the device's electrical properties after neutron irradiation is not due to the influence of interface defects.

Neutrons, as highly penetrating particles, interact with atoms in the β-Ga_2O_3 material, displacing them from their lattice positions and creating vacancies. If the vacancies and interstitial atoms are still influenced by their elastic and Coulomb fields, they may recombine. However, if they exceed the range of these fields, they can form interstitial-vacancy pairs, i.e., Frenkel defects. These defects can lead to device performance degradation or permanent damage. Therefore, the degradation of the device's electrical properties after irradiation may be caused by changes in the bulk defect concentration induced by neutron irradiation.

Figure 6 shows the variation in capacitance of the device before and after irradiation. Through the change in C-V characteristics, it was calculated that the carrier concentration of the device decreased from 8.328×10^{15} cm^{-3} to 7.734×10^{15} cm^{-3} after neutron irradiation. This reduction in carrier concentration is the direct cause of the decrease in the device's forward current.

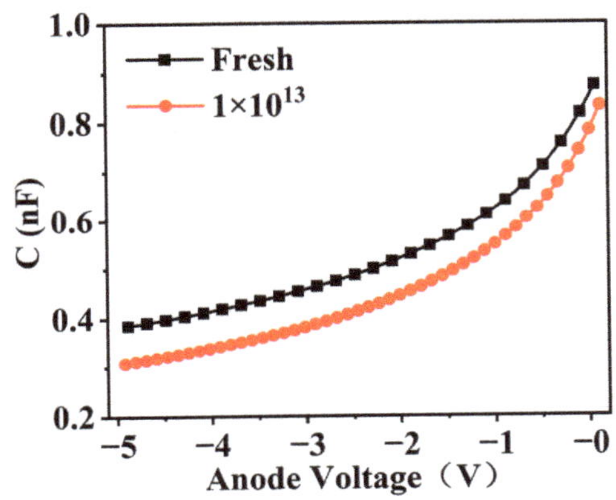

Figure 6. Variation of capacitance of the device before and after neutron irradiation.

The reduction in carrier concentration is attributed to the carrier removal effect of bulk defects. DLTS is a powerful and widely used technique for characterizing deep-level defects in semiconductors. It operates by applying a voltage pulse to the sample, which captures carriers (electrons or holes) into deep-level traps within the material. The subsequent release of these trapped carriers generates a transient current signal that is measured as a function of temperature. By analyzing the temperature dependence of these transient signals, DLTS can provide detailed information on the energy levels and concentrations and capture cross-sections of deep-level defects. This non-destructive and highly sensitive method is essential for understanding the electronic properties of semiconductor materials and devices, particularly in the context of quality control and fundamental research. To verify this, we conducted validation using DLTS defect spectra. During the test, the reverse voltage (U_R) was set at −5 V, the U_P at 10 V, the T_W at 7 s, and the pulse width (T_P) at 1 ms. The test results are shown in Figure 7.

As can be observed from the figure, the temperature at which the defects within the device are excited remains the same after irradiation, indicating that the defect energy levels have not changed. However, the peak value of the vertical axis in the DLTS spectra increased from 1.64 fF to 1.97 fF after irradiation, which suggests an increase in defect concentration. The defect energy levels and concentrations can be determined through the Arrhenius linear relationship plot, as shown in Figure 8. Before irradiation, the primary defect within the device had an energy level depth of Ec−0.28 eV, with a defect concentration of 2.67 × $10^{15}cm^{-3}$. After irradiation, the defect energy level deepened to Ec−0.31 eV, and the defect concentration decreased to 8.94 × 10^{14} cm^{-3}.

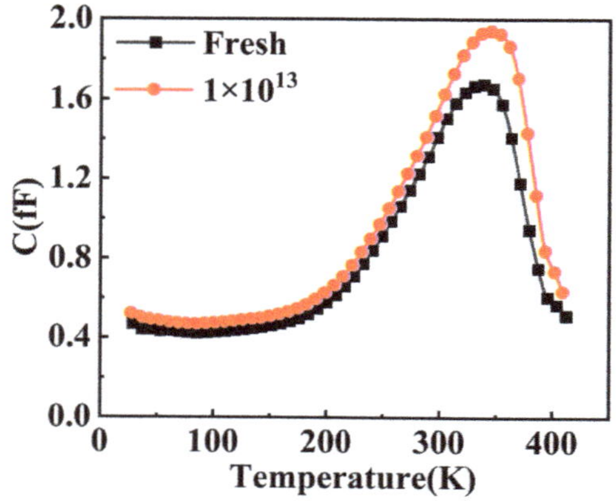

Figure 7. DLTS defect spectra of the device before and after neutron irradiation.

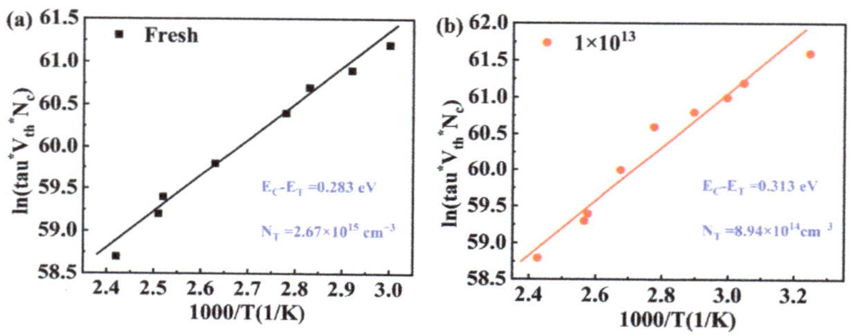

Figure 8. (**a**) Arrhenius plot of the energy levels of the device before irradiation. (**b**) Arrhenius plot of the energy levels of the device after irradiation.

The DLTS spectra unveiled the emergence of specific defect states introduced by neutron irradiation, which function as carrier traps and substantially influence the carrier concentration within the material. This observation corroborates the reduction in carrier concentration following neutron irradiation, thereby affirming that the bulk defects generated by neutron irradiation are the primary cause of the carrier removal effect.

C. TCAD Simulation

TCAD (Technology Computer-Aided Design) simulation is a powerful tool in the semiconductor industry for the development, optimization, and characterization of semiconductor devices and processes. It provides a comprehensive framework for various aspects of semiconductor manufacturing, including the simulation of process steps, device structures, and electrical performance. TCAD tools, such as those from Silvaco and Synopsys, can model complex physical phenomena, optimize device performance, and predict the impact of process variations on device characteristics. These simulations are crucial for understanding the behavior of advanced semiconductor devices. TCAD is also used for device optimization and reliability analysis.

Therefore, to theoretically validate our experimental results, we performed two-dimensional numerical simulations of the SBD devices after neutron irradiation using the Sentaurus 2018 TCAD software. To ensure the accuracy of the simulation results, we employed a variety of physical models, including the Shockley–Read–Hall (SRH) recombination model, high-field saturation model, thermionic emission model, Auger recombination model, and bandgap narrowing model. The material parameters of gallium oxide used in the simulation were derived from previous studies and are specifically summarized in Table 3.

Table 3. β-Ga_2O_3 material parameters used in simulation [27–29].

Parameter	Value (Unit)
Bandgapat 0 K, Eg(0)	4.823 (eV)
Hole Mobility	1.0×10^{-6} (cm^2/V·S)
Electron Mobility	300 (cm^2/V·S)
Dielectric Constant	10.0
Hole Life	2.1×10^{-8} (S)
Electron Life	2.0×10^{-10} (S)
Temperature	300 (K)

After completing the simulation parameter settings, we first conducted a simulation of the TID induced by X-ray irradiation and extracted the relevant physical parameters of the device. The results are shown in Figure 9.

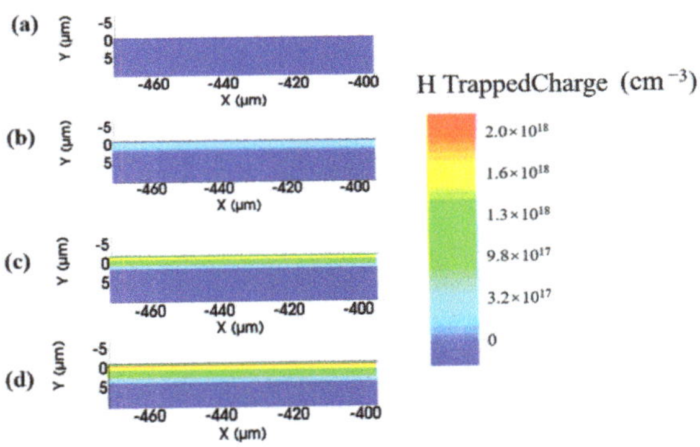

Figure 9. Distribution of hole-trapped charge inside the device after X-ray irradiation at different doses: (**a**) Fresh, (**b**) 500 Krad, (**c**) 1 Mrad, and (**d**) 2 Mrad.

Figure 9 illustrates that the hole density in the device's epitaxial layer increases with higher radiation doses, aligning closely with our experimental findings. The simulation results of the device's current density distribution are shown in Figure 10.

As shown in Figure 10, the simulation results indicate that the internal current density of the device increases with the increase in irradiation dose. In conjunction with the analysis of Figures 2b and 9, this increase is attributed to the higher concentration of interface defects caused by the TID. This finding corroborates our experimental results.

During the neutron irradiation simulation, since the defect energy levels remained almost unchanged before and after irradiation, we added defects with an energy level of Ec−0.3 eV to the device both before and after irradiation. The defect concentration in the device before irradiation was set to 2.67×10^{15} cm^{-3}, and the defect concentration in the device after irradiation was set to 8.94×10^{14} cm^{-3}. This approach allowed us to more accurately simulate the experimental results. The simulation results are shown in Figure 11.

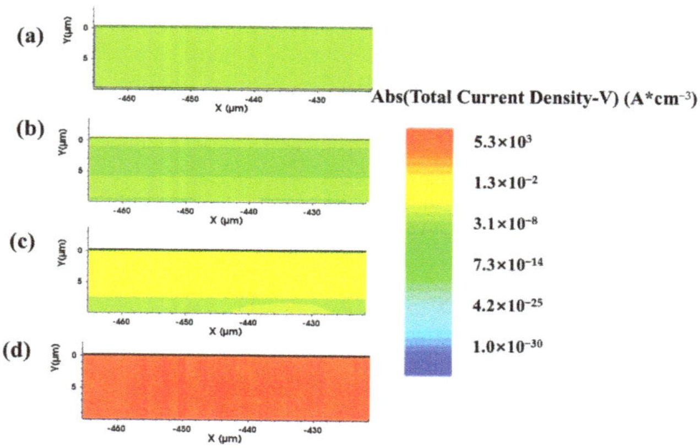

Figure 10. Distribution of total current density inside the device after X-ray irradiation at different doses: (**a**) Fresh, (**b**) 500 Krad, (**c**) 1 Mrad, and (**d**) 2 Mrad.

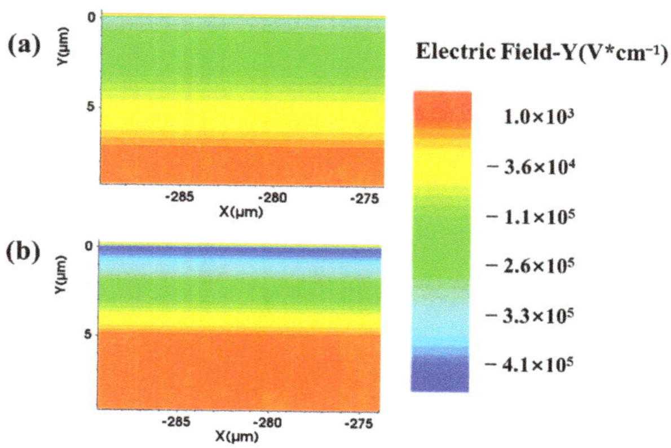

Figure 11. Illustration of the electric field intensity distribution within the epitaxial layer under a reverse bias of 100 V both before and after irradiation: (**a**) pre-irradiation and (**b**) post-irradiation.

As illustrated in Figure 11, after introducing defects into the device, a reverse bias voltage of 100 V was applied. The electric field intensity in the sensitive region of the epitaxial layer post-irradiation was observed to be higher compared to the pre-irradiation state. This finding is consistent with the neutron irradiation results shown in Figure 5b. Neutron irradiation generates a significant number of lattice defects in semiconductor materials. These defects, including vacancies, interstitial atoms, and complex defects, disrupt the periodic structure of the lattice, leading to increased carrier scattering and a consequent reduction in carrier mobility. To further investigate this effect, the hole mobility in the epitaxial layer region was extracted both before and after irradiation, as shown in Figure 12.

As shown in Figure 12, after irradiation, the mobility of holes and other defects decreases, which leads to an increase in the local electric field inside the device. Meanwhile, acting as leakage channels, they enhance the reverse current density, thereby reducing the breakdown voltage of the device and affecting its reliability.

In summary, neutron irradiation introduces a large number of lattice defects, such as vacancies and interstitial atoms, into the semiconductor material of the diode. These defects disrupt the crystal structure of the material, leading to a decrease in carrier mobility and an increase in recombination rates. Since these defects are randomly and disorderly distributed, it is almost impossible to restore the material to its pre-irradiation ordered state. This results in a decline in the forward current transport performance of the diode and an increase in the reverse leakage current. These changes are irreversible because the defects introduced by irradiation permanently alter the carrier transport paths and recombination mechanisms.

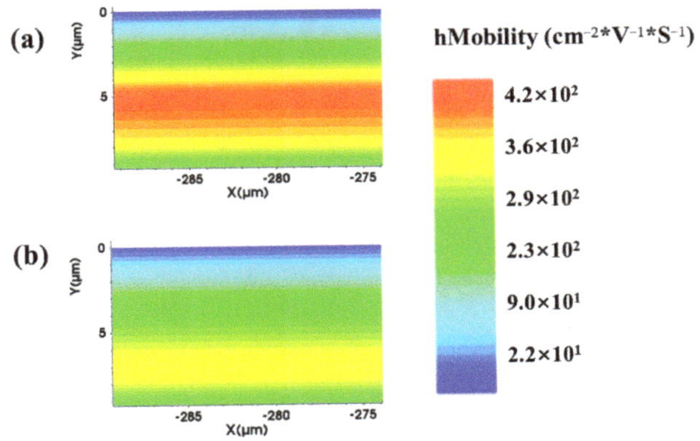

Figure 12. The simulation results of the change in hole mobility in the device before and after neutron irradiation: (**a**) before irradiation and (**b**) after irradiation.

4. Conclusions

This study examines the degradation of electrical properties in β-Ga_2O_3 SBDs due to varying doses of X-ray irradiation using both experimental and simulation methods. The results indicate that the degradation of device performance is associated with interface defects induced by the TID of X-rays. In contrast, neutron irradiation primarily causes displacement damage in the devices. DLTS characterization revealed that the energy levels of bulk defects caused by displacement damage remain unchanged, but their concentration increases. The carrier removal effect induced by these bulk defects reduces the forward current density of the devices after neutron irradiation. Meanwhile, these defects serve as leakage channels, leading to an increase in reverse current density, a decrease in carrier mobility inside the device, and a reduction in the breakdown voltage of the device. By employing the TCAD two-dimensional simulation tool, this study investigates the effects of TID and displacement damage on devices exposed to varying doses of X-ray and neutron irradiation. The simulation results demonstrate strong consistency with experimental observations, validating the accuracy of the computational approach. Furthermore, the findings shed light on the degradation mechanisms affecting the electrical properties of β-Ga_2O_3 under radiation exposure, offering valuable insights for enhancing the radiation tolerance of β-Ga_2O_3 devices. These results provide a foundation for optimizing radiation-hardening processes in future device development.

Author Contributions: Methodology, W.F.; Validation, X.T.; Writing—original draft, P.Z.; Writing—review & editing, T.M. All authors have read and agreed to the published version of the manuscript.

Funding: This research was funded by the fund of the GuangDong Basic and Applied Basic Research Foundation (No. 2022A1515111049) and the National Natural Science Foundation of China (No. 12305299).

Data Availability Statement: The datasets presented in this article are not readily available because the data are part of an ongoing study.

Acknowledgments: We gratefully acknowledge the technical support provided by Xi'an Jiaotong University and Science and Technology on Reliability Physics and Application of Electronic Component Laboratory.

Conflicts of Interest: The authors declare no conflicts of interest.

References

1. Higashiwaki, M.; Sasaki, K.; Kuramata, A.; Masui, T.; Yamakoshi, S. Gallium oxide (Ga_2O_3) metal-semiconductor field-effect transistors on single-crystal β-Ga_2O_3 (010) substrates. *Appl. Phys. Lett.* **2012**, *100*, 013504. [CrossRef]
2. Pearton, S.J.; Yang, J.; Cary, P.H.; Ren, F.; Kim, J.; Tadjer, M.J.; Mastro, M.A. A review of Ga_2O_3 materials, processing, and devices. *Appl. Phys. Rev.* **2018**, *5*, 011301. [CrossRef]
3. Green, A.J.; Chabak, K.D.; Heller, E.R.; Fitch, R.C.; Baldini, M.; Fiedler, A.; Jessen, G.H. β-Ga_2O_3 MOSFETs for radio frequency operation. *IEEE Electron Device Lett.* **2016**, *37*, 902–905. [CrossRef]
4. Wong, M.H.; Sasaki, K.; Kuramata, A.; Yamakoshi, S.; Higashiwaki, M. Field-plated Ga_2O_3 MOSFETs with a breakdown voltage of over 750 V. *IEEE Electron Device Lett.* **2016**, *37*, 212–215. [CrossRef]
5. Zhang, Y.; Joishi, C.; Xia, Z.; Brenner, M.; Lodha, S.; Rajan, S. Demonstration of high mobility and quantum transport in modulation-doped β-$(Al_xGa_{1-x})_2O_3/Ga_2O_3$ heterostructures. *Appl. Phys. Lett.* **2018**, *112*, 173502. [CrossRef]
6. Chen, X.; Ren, F.; Gu, S.; Ye, J. Review of gallium oxide based field-effect transistors and Schottky barrier diodes. *Chin. Phys. B* **2019**, *28*, 017105.
7. Oh, S.; Jung, Y.; Mastro, M.A.; Kim, J.; Ren, F. Radiation effects in Ga_2O_3 materials and devices. *J. Mater. Chem. C* **2020**, *8*, 13054–13065.
8. Ahmadi, E.; Oshima, Y.; Wu, F. Radiation damage effects in Ga_2O_3 materials and devices. *J. Appl. Phys.* **2020**, *127*, 131101.
9. Wu, C.; Wu, F.M.; Hu, H.Z.; Wang, S.L.; Liu, A.P.; Guo, D.Y. Radiation-induced gallium vacancy defects and carrier mobility degradation in β-Ga_2O_3 photodetectors. *Mater. Today Phys.* **2022**, *28*, 100883. [CrossRef]
10. Polyakov, A.Y.; Smirnov, N.B.; Govorkov, A.V.; Pearton, S.J. Fast neutron irradiation effects in n-GaN and β-Ga_2O_3. *J. Vac. Sci. Technol. B* **2007**, *25*, 436–442. [CrossRef]
11. Pearton, S.J.; Ren, F.; Patrick, E.; Law, M.E.; Polyakov, A.Y. Review—Ionizing radiation damage effects on GaN and β-Ga_2O_3 devices. *ECS J. Solid State Sci. Technol.* **2016**, *5*, Q35–Q60.
12. Cai, Z.; He, X.; Wang, K.; Long, H. Enhancing Performance of GaN/Ga_2O_3 p-n Junction UVC Photodetectors via Interdigitated Structure. *Small Methods* **2023**, *7*, 2301148.
13. Higashiwaki, M.; Sasaki, K.; Kuramata, A.; Masui, T.; Yamakoshi, S. Development of gallium oxide power devices. *Phys. Status Solidi A* **2014**, *211*, 21–26.
14. Liu, M.; Hua, M.; Tian, X.; Wang, Z.; Gao, H.; Wang, W.; Chen, Y.; Zhang, C.; Zhao, S.; Feng, Q.; et al. Effect of gamma irradiation on β-Ga_2O_3 vertical Schottky barrier diode. *Appl. Phys. Lett.* **2023**, *123*, 212103. [CrossRef]
15. Holmes-Siedle, A.G.; Adams, L. *Handbook of Radiation Effects*; Oxford University Press: Oxford, UK, 1993.
16. Sheoran, H.; Kumar, V.; Singh, R. A Comprehensive Review on Recent Developments in Ohmic and Schottky Contacts on Ga_2O_3 for Device Applications. *ACS Appl. Electron. Mater.* **2022**, *4*, 2589–2628. [CrossRef]
17. Sze, S.M.; Li, Y.; Ng, K.K. *Physics of Semiconductor Devices*; Nework Wiley: Hoboken, NJ, USA, 1981.
18. Zheng, X.-F.; Dong, S.-S.; Ji, P.; Wang, C.; He, Y.-L.; Lv, L.; Ma, X.-H.; Hao, Y. Characterization of bulk traps and interface states in AlGaN/GaN heterostructure under proton irradiation. *Appl. Phys. Lett.* **2018**, *112*, 233504. [CrossRef]
19. Ai, W.-S.; Liu, J.; Feng, Q.; Zhai, P.-F.; Hu, P.-P.; Zeng, J.; Zhang, S.-X.; Li, Z.-Z.; Liu, L.; Yan, X.-Y.; et al. Degradation of β-Ga_2O_3 Schottky barrier diode under swift heavy ion irradiation. *Chin. Phys. B* **2021**, *30*, 056110. [CrossRef]
20. Vandamme, L.K.J.; Li, X.; Rigaud, D. 1/f noise in MOS devices, mobility fluctuations, and number fluctuations. *IEEE Trans. Electron Devices* **1994**, *41*, 1936–1945.
21. Hooge, F.N. 1/f noise sources. *IEEE Trans. Electron Devices* **1994**, *41*, 1926–1935.
22. Simoen, E.; Claeys, C. On the flicker noise in submicron silicon MOSFETs. *Solid-State Electron.* **1999**, *43*, 865–882.
23. Balandin, A.A. Low-frequency 1/f noise in graphene devices. *Nat. Nanotechnol.* **2013**, *8*, 549–555. [PubMed]
24. Fleetwood, D.M. Radiation effects in a fluctuating world: The role of noise in semiconductor devices and circuits. *IEEE Trans. Nucl. Sci.* **2018**, *65*, 1465–1481.
25. Jantsch, O. Flicker (1/f) noise generated by a random walk of electrons in interfaces. *IEEE Trans. Electron Devices* **1987**, *34*, 1100–1115. [CrossRef]
26. Yue, S.; Zheng, X.; Hong, Y.; Zhang, X.; Zhang, F.; Wang, Y.; Ma, X.; Hao, Y. Effects of Neutron Irradiation on Electrical Performance of β-Ga_2O_3 Schottky Barrier Diodes. *IEEE Trans. Electron Devices* **2023**, *70*, 3026–3030. [CrossRef]
27. Ma, H.; Wang, W.; Cai, Y.; Wang, Z.; Zhang, T.; Feng, Q.; Chen, Y.; Zhang, C.; Zhang, J.; Hao, Y. Analysis of single event effects by heavy ion irradiation of Ga_2O_3 metal–oxide–semiconductor field-effect transistors. *J. Appl. Phys.* **2023**, *133*, 084501. [CrossRef]

28. Wong, H.Y.J. TCAD simulation models, parameters, and methodologies for β-Ga_2O_3 power devices. *ECS J. Solid State Sci. Technol.* **2023**, *12*, 045001.
29. Zhou, H.; Zeng, S.; Zhang, J.; Liu, Z.; Feng, Q.; Xu, S.; Zhang, J.; Hao, Y. Comprehensive study and optimization of implementing p-NiO in β-Ga_2O_3 based diodes via TCAD simulation. *Crystals* **2021**, *11*, 1186. [CrossRef]

Disclaimer/Publisher's Note: The statements, opinions and data contained in all publications are solely those of the individual author(s) and contributor(s) and not of MDPI and/or the editor(s). MDPI and/or the editor(s) disclaim responsibility for any injury to people or property resulting from any ideas, methods, instructions or products referred to in the content.

electronics

MDPI

Article

Total Ionizing Dose Effects on Lifetime of NMOSFETs Due to Hot Carrier-Induced Stress

Yujuan He [1,2], Rui Gao [2], Teng Ma [2], Xiaowen Zhang [2], Xianyu Zhang [2,3] and Yintang Yang [1,*]

[1] School of Microelectronics, Xidian University, Xi'an 710071, China; heyujuan@ceprei.com
[2] Science and Technology on Reliability Physics and Application of Electronic Component Laboratory, China Electronic Product Reliability and Environmental Testing Research Institute, Guangzhou 511370, China; 2222031324@s.xmut.edu.cn (X.Z.)
[3] School of Optoelectronics and Communication Engineering, Xiamen University of Technology, Xiamen 361005, China
* Correspondence: ytyang@xidian.edu.cn

Abstract: This study systematically investigates the mechanism by which total ionizing dose (TID) affects the lifetime degradation of NMOS devices induced by hot-carrier injection (HCI). Experiments involved Cobalt-60 (Co-60) gamma-ray irradiation to a cumulative dose of 500 krad (Si), followed by 168 h annealing at 100 °C to simulate long-term stability. However, under HCI stress conditions (V_D = 2.7 V, V_G = 1.8 V), irradiated devices show a 6.93% increase in threshold voltage shift (ΔV_{th}) compared to non-irradiated counterparts. According to the IEC 62416 standard, the lifetime degradation of irradiated devices induced by HCI stress is only 65% of that of non-irradiated devices. Conversely, when the saturation drain current (I_{Dsat}) degrades by 10%, the lifetime doubles compared to non-irradiated counterparts. Mechanistic analysis demonstrates that partial neutralization of E′ center positive charges at the gate oxide interface by hot electrons weakens the electric field shielding effect, accelerating ΔV_{th} drift, while interface trap charges contribute minimally to degradation due to annealing-induced self-healing. The saturation drain current shift degradation primarily correlates with electron mobility variations. This work elucidates the multi-physics mechanisms through which TID impacts device reliability and provides critical insights for radiation-hardened design optimization.

Keywords: total irradiation dose; hot-carrier injection; lifetime

1. Introduction

The rapid advancement of space science and aerospace technology has transformed deep space exploration from science fiction into an operational reality, with sustained Mars missions now exceeding 1500 sols and lunar base construction underway. While public attention often focuses on spectacular achievements like extraterrestrial landings, the engineering community faces critical reliability challenges in semiconductor devices that power these ambitious missions. Space radiation environments induce total ionizing dose effects that generate oxide trap charges and interface trap charges in MOS devices, causing threshold voltage drift and leakage current increases exceeding several orders of magnitude [1,2]. Concurrently, prolonged operational stresses during active mission phases subject devices to hot-carrier injection mechanisms, where electric fields > 6 MV/cm accelerate charge carriers, creating additional interface traps that degrade transconductance by 18–22% [3–5]. These parallel degradation pathways demonstrate striking similarities in their physical manifestations, yet current reliability assessments persist in evaluating them through isolated single-mechanism approaches.

Emerging evidence suggests synergistic interactions between radiation-induced defects and electrothermal stress mechanisms [6,7]. Accelerated aging tests reveal non-linear degradation patterns where combined TID-HCI exposure causes 40% greater threshold voltage shifts than linear superposition predictions, particularly in advanced 14 nm FinFET nodes. This interaction stems from radiation-created defect precursors lowering activation energy for subsequent hot-carrier damage by 0.15 eV, while trapped charges modify local electric fields to enhance carrier injection probabilities sixfold. Such cross-mechanism amplification effects remain unaddressed in current space qualification standards, potentially leading to 55–70% lifetime overestimations for five-year Mars missions.

Some combined studies on space radiation and electrical stress reliability in semiconductor devices have emerged globally [8–15]. References [8,9] conduct preliminary investigations on the combined damage of total ionizing dose radiation and hot-carrier injection in MOS capacitors, concluding that their synergistic damage effect does not follow simple linear superposition, though the underlying causes and mechanisms are not analyzed. Studies [10,11] have investigated the impact of different TID testing methodologies—such as direct HCI testing post-TID irradiation, or TID irradiation followed by high-temperature annealing and subsequent HCI testing—on the interplay between TID effects and HCI-induced degradation. Reference [12] examines TID radiation effects on hot-carrier behavior in 10 nm FinFET devices. Study [13] proposes TID-induced annealing effects in the buried oxide layer of SOI devices under HCI-induced stress, and Reference [14] reveals differential impacts of TID-induced charges on hot-carrier effects in *n*-channel versus *p*-channel FinFETs. Additionally, Reference [15] identifies layout-dependent radiation-enhanced hot-carrier degradation in 65 nm MOS devices. However, these studies rarely analyze the impact of the combined stress of total ionizing dose irradiation and hot carriers on device degradation lifetime.

This study investigates the impact of TID irradiation on the hot-carrier lifetime of NMOS devices. Co-60 radiation is employed to irradiate the devices with a total dose of 500 krad (Si), followed by annealing at 100 °C for 168 h. HCI stress tests under three voltage conditions for 5000 s are conducted on both irradiated and non-irradiated devices to analyze their characteristic degradation and defect evolution under combined TID and HCI stress. According to the IEC 62416 standard for device lifetime calculation under hot-carrier injection effects, the lifetime under HCI stress associated with a 10% increase in threshold voltage for irradiated devices is approximately 65% of that for non-irradiated devices. Conversely, when the saturation leakage current decreases by 10% under HCI stress, the lifetime of irradiated devices is approximately twice that of non-irradiated devices.

2. Materials and Methods

The tested devices are Bulk N-Metal-Oxide-Semiconductor (NMOS) transistors fabricated by the Semiconductor Manufacturing International Corporation (SMIC), featuring a gate oxide thickness of 2 nm, a gate length of 130 nm, and a width-to-length ratio (W/L) of 100:1.

The total dose irradiation test is conducted with a dose rate of 50 rad (Si)/s and a total radiation dose of 500 krad (Si), under ON bias conditions (gate voltage V_G = 1.2 V, drain voltage, source voltage, and substrate voltage V_D = V_S = V_{sub}= 0 V). Irradiation is performed at the Xinjiang Technical Institute of Physics and Chemistry, Chinese Academy of Sciences with a Cobalt-60 source. Following the total dose irradiation, an annealing test is conducted at a high temperature (100 °C) under ON bias for a duration of 168 h. Given that the actual space environment exhibits an extremely low dose rate (0.01 rad (Si)/s), high-temperature annealing tests following total ionizing dose irradiation are adopted to simulate space radiation effects. This approach ensures that the combined TID and HCI stress testing results more accurately reflect realistic space operating conditions.

According to the IEC 62416-2010 standard [16], the drain voltage for hot-carrier testing is determined based on the device's drain breakdown voltage. The gate voltage is specified by applying the maximum substrate current method at each corresponding drain bias. Both the source and the substrate are connected to the ground. The test conditions for the hot-carrier test are presented in Table 1, and the test is performed under room temperature conditions (25 °C ± 2 °C).

Table 1. The test condition of samples.

HCI Stress Condition	V_D/V	V_G/V	V_S/V	V_{sub}/V
Condition 1	2.7	1.8	0	0
Condition 2	2.5	1.7	0	0
Condition 3	2.3	1.6	0	0

The hot-carrier test conditions for samples subjected to total dose rate irradiation, post-irradiation annealing, and non-irradiated samples are identical. For each sample under each test condition, five devices are subjected to the hot-carrier test, with a test duration of 5000 s. The *I-V* characteristics of the NMOS devices are measured using the Keysight 1500 A Semiconductor Device Parameter Analyzer (Keysight Technologies, Santa Rosa, CA, USA). The Keysight 1500 A offers voltage measurement accuracy at the 1 μV level and current measurement accuracy at the 1fA level.

3. Results and Discussion

3.1. Irradiation and Annealing Test Results

The transfer characteristic curves (I_D-V_G curves) before and after total dose irradiation and annealing tests are shown in Figure 1.

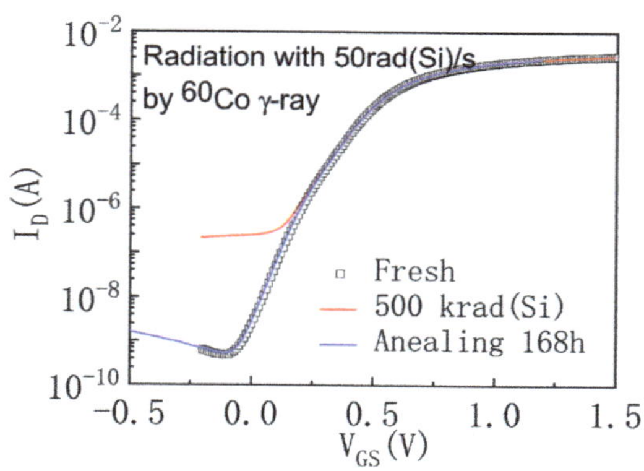

Figure 1. Transfer characteristic curves (I_D-V_G curves) before and after irradiation and annealing.

Experimental results demonstrated distinct radiation effects and recovery behaviors in NMOS devices: Under a total ionizing dose of 500 krad (Si), the threshold voltage exhibited exceptional stability with a drift of <5 mV, confirming minimal oxide charge trapping influence on gate control. However, off-state leakage current surged from 10^{-9} A to 10^{-7} A due to radiation-induced interface traps creating parasitic conduction paths. Subsequent annealing at 100 °C for 168 h fully restored leakage to 10^{-9} A and recovered the I_D-V_G curve to its pre-irradiation profile, attributed to hydrogen-mediated passivation of interface defects during thermal treatment. These findings establish that while ionizing radiation primarily degrades off-state leakage through trap-assisted tunneling, standard annealing protocols can completely reverse such damage without compromising threshold voltage integrity, highlighting the robustness of NMOS devices in radiation-aware applications.

To spread device-to-device variation effects on irradiation and annealing responses, threshold voltage, maximum transconductance ($G_{m\text{-}max}$), and subthreshold swing (SS) distributions are statistically analyzed across multiple devices, as shown in Figure 2. Notably, the I_D range of 10^{-5} A to 10^{-4} A is selectively used for SS extraction to eliminate miscalculations caused by radiation-elevated off-state leakage currents. The threshold voltage shifted from 0.4745 V (pre-irradiation) to 0.4771 V after 500 krad irradiation, only resulting in a ΔV_{th} of +2.6 mV. This positive drift dominance indicates that interface trap charges (contributing to V_{th} increase) outweighed oxide-trapped charges (inducing V_{th} decrease). Post-annealing V_{th} remained at 0.4768 V, showing no significant recovery trend. Further analysis of SS demonstrated irradiation-induced degradation from 0.1420 V/decade to 0.1433 V/decade, consistent with interface trap generation, followed by near-complete restoration to 0.1421 V/decade after annealing. The change in the maximum transconductance value is also very small during the total dose irradiation process, and the change in the maximum transconductance value after irradiation is less than 1 mS.

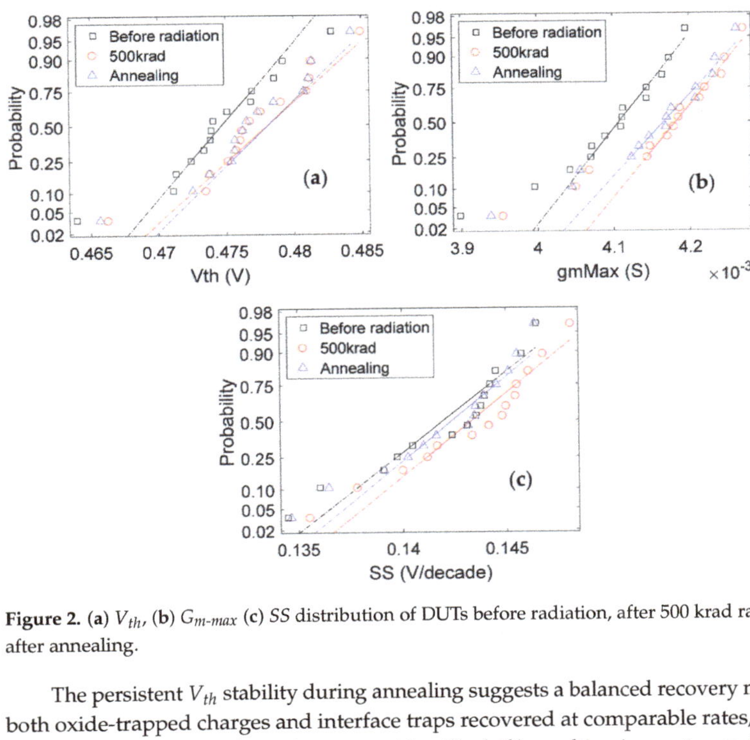

Figure 2. (**a**) V_{th}, (**b**) $G_{m\text{-}max}$ (**c**) SS distribution of DUTs before radiation, after 500 krad radiation and after annealing.

The persistent V_{th} stability during annealing suggests a balanced recovery mechanism: both oxide-trapped charges and interface traps recovered at comparable rates, with their opposing contributions (negative vs. positive V_{th} drift) resulting in a net-zero drift. These observations confirm that while interface traps dominate SS degradation and partial V_{th} shifts, the counteracting recovery dynamics between oxide and interface charges maintain V_{th} homeostasis during thermal treatment.

3.2. Hot-Carrier Test Results

After conducting irradiation and annealing tests, hot-carrier tests are performed. The I_D-V_G curves under different time conditions are shown in Figure 3. The results reveal that during the hot-carrier test, the threshold voltage exhibited a positive shift, and the slope of the subthreshold region decreased, indicating an increase in the subthreshold swing.

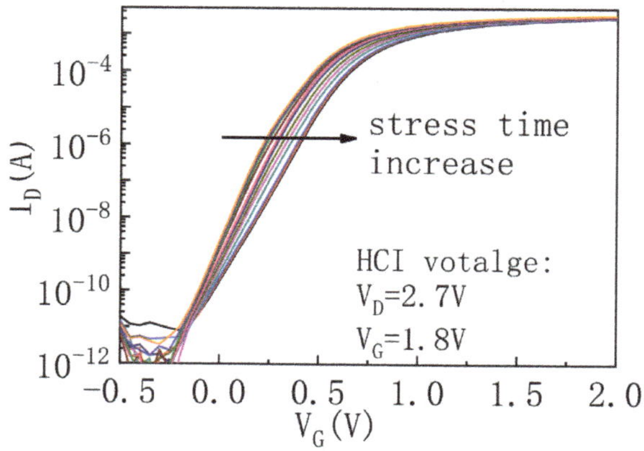

Figure 3. I_D-V_G curves of sample with HCI test.

The V_{th} variations during irradiation, annealing, and hot-carrier tests are illustrated in Figure 4. The results show that the V_{th} of the NMOS device is insensitive to TID irradiation, with minimal changes induced by irradiation ($\Delta V_{th} \approx 2.6$ mV, as shown in Figure 2a). However, during the hot-carrier test, the threshold voltage increased rapidly. At HCI condition 1, the V_{th} rose from 0.4771 V to 0.6598 V after 10,000 s of hot-carrier stress, resulting in a ΔV_{th} = 182.7 mV. This shift is significantly higher than the average ΔV_{th} caused by TID irradiation (2.6 mV), highlighting the severe degradation under hot-carrier stress compared to irradiation effects.

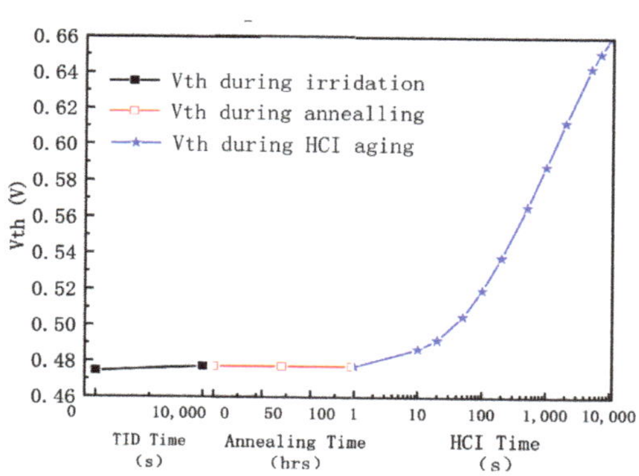

Figure 4. Threshold voltage during irradiation, annealing, and HCI tests.

Hot-carrier injection tests are performed on both non-irradiated and irradiated devices under identical conditions. Figure 5 illustrates the average values and variances of threshold voltage shift (ΔV_{th}) and saturation drain current shift (ΔI_{Dsat}) for five devices under three stress conditions as functions of testing time under hot-carrier stress. ΔV_{th} represents the change in threshold voltage between before and after HCI testing, while ΔI_{Dsat} denotes the change in saturation drain current between before and after HCI testing. Observations from Figure 5 reveal that the degradation rate of ΔV_{th} in total-dose-irradiated NMOS devices under HCI stress is slightly faster than that of non-irradiated devices, whereas the degradation rates of ΔI_{Dsat} under HCI stress are slower in irradiated devices compared to non-irradiated ones, exhibiting trends opposite to ΔV_{th}. As shown in Figure 5a, under HCI Stress Condition 1 (V_D = 2.7 V, V_G = 1.8 V), the ΔV_{th} of irradiated devices increases by 185 mV after 5000 s of HCI stress, while non-irradiated devices show a ΔV_{th} of 173 mV, with a minimal difference of 12 mV. Under HCI Stress Condition 3 (V_D = 2.3 V, V_G = 1.6 V), the ΔV_{th} values after 5000 s of stress are 71 mV (irradiated) and 68 mV (non-irradiated), differing by only 3 mV. This indicates that the ΔV_{th} difference between irradiated and non-irradiated devices is negligible under low HCI stress but becomes more pronounced as stress intensity increases. In contrast, ΔI_{Dsat} exhibits an opposing trend to ΔV_{th}. From Figure 5b, under HCI Stress Condition 1, ΔI_{Dsat} reaches −2.11 mA (irradiated) and −2.34 mA (non-irradiated) after 5000 s of stress, with a difference of +0.23 mA. Under HCI Stress Condition 3, ΔI_{Dsat} values are −0.79 mA (irradiated) and −0.95 mA (non-irradiated), differing by +0.16 mA. It is evident from Figure 5b that the saturation leakage current decreases with increasing HCI stress time. However, the reduction in saturation leakage current for the unirradiated device is less pronounced compared to the irradiated device, which is contrary to the variation trend of ΔV_{th}. Furthermore, the difference in ΔI_{Dsat} between the irradiated and unirradiated devices increases slightly as the hot-carrier stress voltage rises.

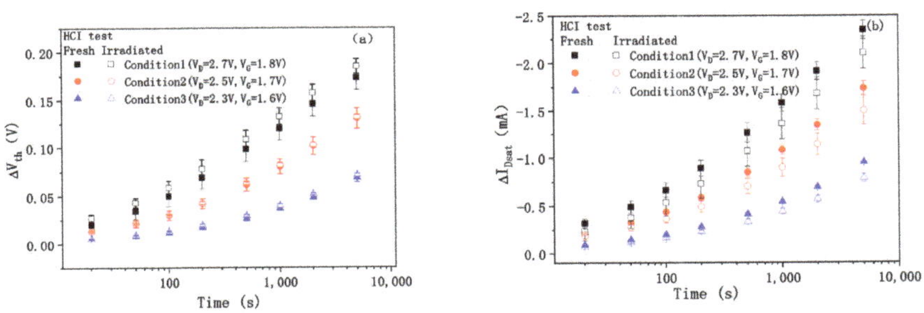

Figure 5. Variations in NMOSFET parameter drift under three hot-carrier injection stress conditions with stress duration: (**a**) threshold voltage shift and (**b**) saturation drain current shift.

3.3. Comparative Analysis of Hot-Carrier Lifetime

To quantitatively evaluate the impact of irradiation on hot-carrier degradation, degradation model parameters extracted before and after irradiation are compared.

According to IEC 62416-2010, the relationship between the degradation amount of MOS device parameters and the degradation time, as well as the stress voltage value, during hot-carrier injection effect testing can be expressed by the following formula:

$$\Delta D = A \times \exp(B/V_{\text{d,stress}}) \times t^{C} \tag{1}$$

where ΔD represents the degradation amount of the MOS device parameter caused by hot-carrier injection stress, A and B are coefficients determined through the same global fitting procedure, C is the degradation time exponent, and t is time. Based on Formula (1), the threshold voltage shift caused by hot carriers can be expressed by the following equation:

$$\Delta V_{\text{th}} = A \times \exp(B/V_{\text{d,stress}}) \times t^{C} \tag{2}$$

Using the least squares method, a global fitting is performed based on Equation (1) for the degradation data for unirradiated devices under three stress voltages ($V_{D,stress}$ = 2.7 V, 2.5 V, 2.3 V), yielding the degradation time exponent $C = 0.42$ in Equation (2), as shown in Figure 6a.

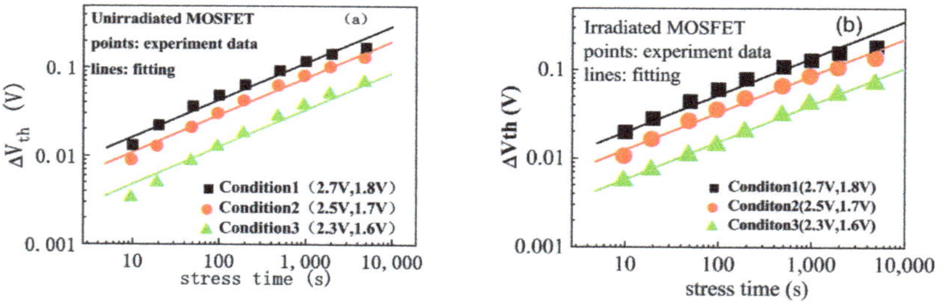

Figure 6. Threshold voltage shifts in the NMOSFET devices measured and predicted from the power law model (**a**) before and (**b**) after irradiation.

By setting the stress time t = 100 s, the threshold voltage shift ΔV_{th} at 1 s corresponds to the value of A × exp (B/$V_{D,stress}$). Based on the linear relationship ΔV_{th} @100 s~1/$V_{D,stress}$, the coefficients are determined as $A = 8.003$ and $B = -19.21$. The hot-carrier degradation model for non-irradiated devices is expressed as follows:

$$\Delta V_{\text{th}} = 8.003 \times \exp(-19.21/V_{\text{D,stress}}) \times t^{0.42} \tag{3}$$

Using the same model parameter extraction methodology, the hot-carrier degradation model for irradiated devices is derived as follows, with the parameter extraction process illustrated in Figure 6b:

$$\Delta V_{\text{th}} = 9.455 \times \exp(-19.20/V_{\text{D,stress}}) \times t^{0.42} \tag{4}$$

According to the IEC 62416-2010 standard [16], based on the model Equations (3) and (4), using ΔV_{th} = 50 mV (approximately 10% degradation of the initial V_{th}) as the failure criterion, the calculated hot-carrier lifetimes for non-irradiated and irradiated devices operating at V_D = 1.2 V are 2.018 × 10^{11} s and 1.330 × 10^{11} s, respectively. This indicates that the hot-carrier lifetime for 10% threshold voltage degradation in irradiated devices is approximately 65% of that in non-irradiated devices, demonstrating that total ionizing dose radiation effects accelerate the degradation of threshold voltage under hot-carrier stress.

The saturation drain current variation ΔI_{Dsat} can be expressed by the following equation:

$$\Delta I_{\text{Dsat}} = A \times \exp(B/V_{\text{d,stress}}) \times t^{C} \tag{5}$$

By fitting the average ΔI_{Dsat} values of irradiated and non-irradiated devices (as shown in Figure 7), the hot-carrier degradation model for ΔI_{Dsat} in non-irradiated devices is derived as follows:

$$\Delta I_{\text{Dsat}} = 0.0655 \times \exp(-17.776/V_{\text{d,stress}}) \times t^{0.42} \tag{6}$$

Using the same parameter extraction method, the hot-carrier degradation model for ΔI_{Dsat} in irradiated devices is obtained as follows:

$$\Delta I_{\text{Dsat}} = 0.0589 \times \exp(17.998/V_{\text{d,stress}}) \times t^{0.42} \tag{7}$$

Based on Equations (5) and (6), with ΔI_{Dsat} = 0.6 mA (approximately 10% degradation of the initial I_{Dsat} as the failure criterion), the calculated hot-carrier lifetimes for 10% saturation drain current degradation in non-irradiated and irradiated devices operating at V_D = 1.2 V are 2.917 $\times$ 10^{10} s and 5.835 $\times$ 10^{10} s, respectively. This reveals that the hot-carrier lifetime for 10%, I_{Dsat} degradation in irradiated devices is approximately twice that of non-irradiated devices, which is completely contrary to the degradation trend observed for the 50 mV threshold voltage criterion. It can be observed that, despite the minimal threshold voltage shift in the device caused by total dose irradiation, the lifetime corresponding to a 10% saturation leakage current degradation is significantly extended due to the total dose effect.

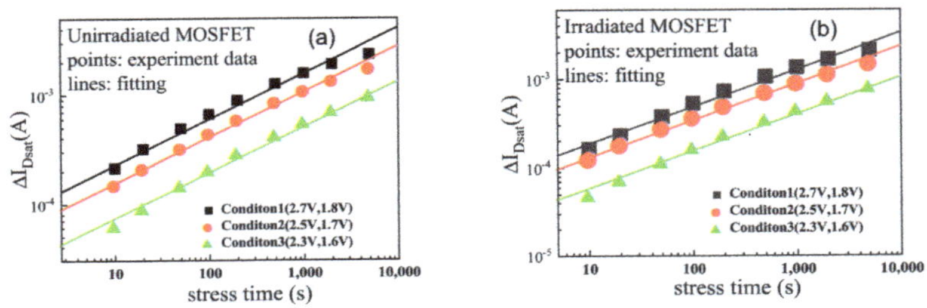

Figure 7. Saturation drain current shifts in the NMOSFET devices measured and predicted from the power law model (**a**) before and (**b**) after irradiation.

3.4. Analysis of Hot-Carrier Lifetime Differences

During TID irradiation of NMOS devices, the structural degradation of the gate oxide layer near the Si-SiO_2 interface initiated with the dissociation of Si-Si bonds, creating dangling bonds that preferentially trapped holes to form E′ centers (Figure 8a). These E′ centers functioned as stable positive oxide trap charges due to their unpaired electrons localized at oxygen vacancy sites. Subsequent hot-carrier injection testing revealed a dynamic interplay between these trapped holes and injected hot electrons: the energetic electrons neutralized the trapped holes in dangling bonds, leading to a net reduction in positive charge within the oxide layer. This charge imbalance directly contributed to a measurable positive shift in threshold voltage, reflecting the cumulative impact of oxide trap charge modulation on device electrostatics.

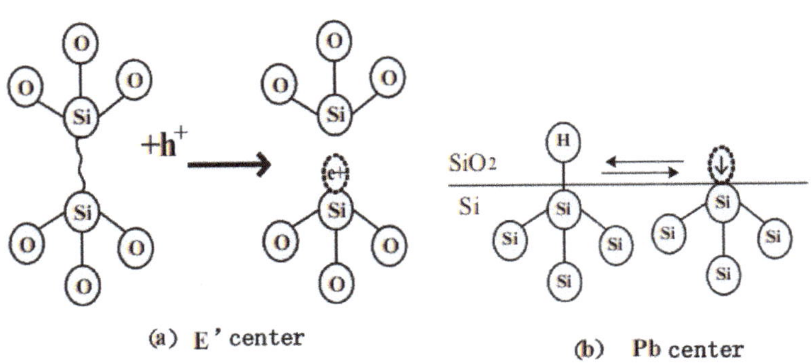

Figure 8. The trap centers: (**a**) E′ center; (**b**) Pb center.

Concurrently, TID irradiation triggered the transformation of silicon dangling bonds at the Si-SiO_2 interface into Pb centers (Figure 8b), which exhibited amphoteric trapping behavior. These centers preferentially captured electrons under operational bias, generating negative interface trap charges. Intriguingly, the radiation-induced oxide trap charges displayed non-monotonic evolution during HCI testing, characterized by partial annihilation of trapped holes followed by charge inversion to negative polarity. Similarly, the interface trap charge density exhibited a bimodal trajectory: initial reduction due to recombination of electron–hole pairs with oxide traps near the interface during early HCI stages, succeeded by progressive accumulation as source-injected electrons became trapped by interface defects. The synergistic effect induced by TID and HCI stresses led to dual-phase evolution of oxide traps and interface traps, where the negative oxide trap charges and negative interface trap charges collectively amplified the positive threshold voltage shift through electrostatic coupling across the gate dielectric.

To elucidate the distinct contributions of oxide trap charges (Q_{ot}) and interface trap charges (D_{it}), comparative analysis of subthreshold swing (*SS*) degradation kinetics under hot-carrier stress is conducted on irradiated and non-irradiated devices (Figure 9). The subthreshold current can be expressed by the following [17]:

$$I_D = I_{D0} \exp \frac{q\psi_s}{nkT} \left\{ 1 - \exp\left(-\frac{qV_D}{kT} \right) \right\} \tag{8}$$

where V_G is the gate–source voltage, V_D is the drain–source voltage, I_{D0} is the channel current when V_G is equal to V_{th}, ψ_s is the surface potential, and n is the body factor which can be expressed by the following:

$$n = \frac{C_{ox} + C_d + C_{it}}{C_{ox}} \tag{9}$$

where C_{ox} is the gate oxide capacitance per unit area, C_d is the depletion capacitance per unit area, and C_{it} is the interface trap capacitance per unit area.

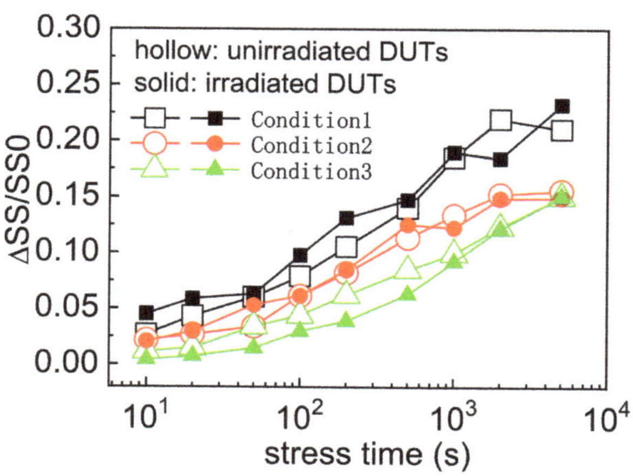

Figure 9. Δ*SS*/*SS0* of HCI tests between irradiated and unirradiated NMOSFET.

So, the subthreshold swing can be expressed by the following:

$$SS \equiv \ln 10 * \frac{dV_G}{d(\ln I_D)} \tag{10}$$

By substituting Equations (7) and (8) into Equation (9) for calculation, the subthreshold swing (SS) can be expressed as follows:

$$SS = \frac{kT}{q} \ln 10 * (1 + \frac{C_d + C_{it}}{C_{ox}}) \tag{11}$$

Since $C_{it} = q^2 \times D_{it}$, where D_{it} is interface trap density, it can be considered that the SS is a function of the interface trap density according to Equation (11).

However, Figure 9 demonstrates negligible differences in SS degradation patterns between irradiated and pristine devices, accompanied by minimal variations in D_{it} evolution during HCI testing. This critical observation conclusively demonstrates that the pronounced hot-carrier lifetime reduction caused by TID radiation predominantly stems from oxide trap charge dynamics (Q_{ot}) rather than interface trap charge accumulation. The mechanistic dominance of Q_{ot} is further corroborated by the charge inversion phenomena and sustained negative charge buildup in the oxide layer, which directly modulates carrier transport and enhances local electric fields, thereby accelerating device degradation under hot-carrier stress conditions.

The calculation formula for the saturation drain current is as follows [17]:

$$I_{Dsat} = \frac{1}{2} \mu_n C_{ox} \frac{W}{L} (V_G - V_{th})^2 \tag{12}$$

where μ_n is electron mobility; W/L is the width-to-length ratio.

Although the total dose irradiation causes only a minor threshold voltage shift, radiation induces positive oxide trap charge and negative interface trap charge. To separate these two effects, the flat band voltage method [18] was employed. The threshold voltage shift (ΔV_{ot}) caused by the oxide trap charge and the threshold voltage shift (ΔV_{it}) caused by the interface trap charge are presented in Figure 10. As can be observed from the figure, despite the total dose irradiation resulting in a small threshold voltage shift of 2.6 mV, ΔV_{ot} is −0.131 V and ΔV_{it} is 0.134 V. This indicates that the induced oxide trap charge and interface trap charge counteract each other in terms of voltage.

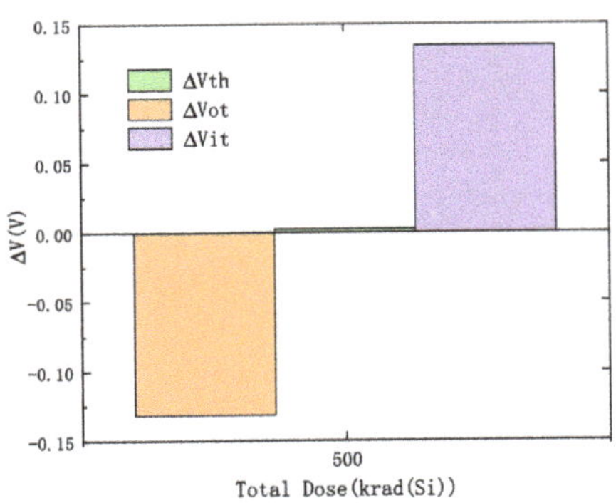

Figure 10. The values of ΔV_{th}, ΔV_{ot}, and ΔV_{it} after irradiation with a total dose of 500 krad (Si).

The formation of oxide trap charges within the gate dielectric layer under total ionizing dose irradiation enhances Coulomb scattering events between charge carriers and lattice defects, significantly increasing the collision probability and degrading electron mobility through two mechanisms: (1) direct momentum transfer via charged defect scattering and (2) modification of the inversion layer charge density. However, during prolonged HCI testing, the injected high-energy electrons penetrate the oxide layer and undergo recombination with trapped holes in radiation-induced defects, partially neutralizing the positive oxide trap charges and even inducing charge inversion at certain defect sites where trapped holes are replaced by electrons. This dynamic charge compensation process reduces the net positive oxide trap charge density in irradiated devices to levels marginally below those observed in pristine devices under identical stress conditions, paradoxically resulting in less severe electron mobility degradation within the irradiated oxide layer.

Since electron mobility varies less in irradiated devices compared to unirradiated devices, this also results in a similar pattern for the maximum transconductance ($G_{m\text{-}max}$). Figure 11 shows the variation in the ratio between the change in subthreshold maximum transconductance ($\Delta G_{m\text{-}max}$) and the pre-stress maximum transconductance ($G_{m\text{-}max0}$) with stress time under three hot-carrier injection stress conditions. It can be seen from the figure that all three HCI stress conditions exhibit a consistent trend: the $\Delta G_{m\text{-}max}/G_{m\text{-}max0}$ value for irradiated devices is smaller than that for unirradiated devices.

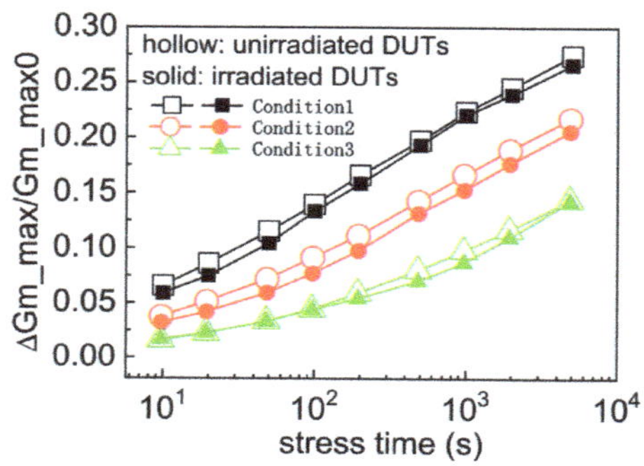

Figure 11. $\Delta G_{m\text{-}max}/G_{m\text{-}max0}$ of HCI tests between irradiated and unirradiated NMOSFET.

The maximum transconductance in the subthreshold region is expressed as follows [17]:

$$G_{\mathrm{m-max}} = \frac{\partial I_D}{\partial V_G} \propto \mu_{\mathrm{n}} C_{\mathrm{OX}} \frac{\mathrm{W}}{\mathrm{L}} V_D \quad (13)$$

As shown in Equation (13), when V_D is fixed, $G_{m\text{-}max}$ depends linearly on electron mobility. Due to mobility degradation in irradiated devices, unirradiated devices exhibit lower ΔI_{Dsat} values compared to irradiated ones. This behavior contrasts with the ΔV_{th} trend, ultimately leading to a significantly extended lifetime of irradiated devices in hot-carrier stress tests compared to unirradiated counterparts.

Consequently, irradiated NMOS devices exhibit smaller electron mobility variations ($\Delta\mu_n$) in the gate oxide region compared to non-irradiated devices during HCI stress, as the radiation-modified defect population provides both trap sites for charge neutralization and recombination centers that mitigate scattering effects. This counterintuitive behavior directly correlates with the observed degradation patterns in ΔI_{Dsat}, confirming that I_{Dsat} degradation kinetics are predominantly governed by electron mobility changes rather than threshold voltage shifts under these coupled radiation–HCI stress conditions. The experimental validation of this mechanism through parameter extraction and defect spectroscopy aligns with theoretical models describing defect–charge interaction dynamics in irradiated MOS systems, establishing a comprehensive framework for understanding reliability degradation in radiation-hardened devices subjected to combined environmental stresses.

4. Conclusions

This study reveals the dual-effect mechanism of total ionizing dose (TID) radiation on HCI reliability in NMOS devices. Experimental data demonstrate accelerated threshold voltage shift in irradiated devices under specific HCI stress conditions, while the degradation of saturation drain current shows a deceleration trend. Quantification using the IEC 62416-2010-based degradation model indicates that when ΔV_{th} = 50 mV is set as the failure criterion, the HCI lifetime of irradiated devices (1.330 $\times$ 10^{11} s) decreases to 65% of non-irradiated counterparts. Conversely, with ΔI_{Dsat} = 0.6 mA as the criterion, the HCI lifetime of irradiated devices (5.835 $\times$ 10^{10} s) doubles compared to non-irradiated ones.

This paradoxical behavior stems from fundamentally distinct physical mechanisms activated by total ionizing dose (TID) radiation: During hot-carrier injection (HCI) stress, the partial neutralization (30–40% recombination efficiency) of positive charges trapped in E′ centers—created by radiation-induced rupture of strained Si-Si bonds at the oxide interface—disrupts the localized electric field screening previously maintained by these trapped holes. The resultant electric field intensification near the gate edge accelerates high-energy carrier injection into the oxide, forming positive feedback that amplifies threshold voltage drift through two interrelated pathways—enhanced trap-assisted tunneling and increased charge trapping probability at radiation-modified defect sites. Conversely, interface traps exhibit diminished contributions to degradation due to thermally activated self-recovery mechanisms during HCI stress, where elevated temperatures facilitate defect annealing and recombination of trapped charges, effectively "healing" a portion of radiation-induced interface states. This annealing-driven recovery suppresses interface trap charge accumulation, decoupling ΔI_{Dsat} degradation from threshold voltage shifts and instead tethering saturation current reduction primarily to electron mobility degradation governed by scattering dynamics. Crucially, the mobility degradation demonstrates an inverse proportionality to oxide trap charge neutralization efficiency—as trapped holes in E′ centers are progressively neutralized by hot electrons, the resultant reduction in Coulomb scattering centers temporarily mitigates electron mobility loss despite the ongoing ΔV_{th} drift. These dual mechanisms collectively reveal a competitive interplay between oxide trap charge modulation and electron mobility driven transport degradation, establishing a predictive model where ΔV_{th} scales with trapped charge neutralization kinetics while ΔI_{Dsat} correlates with defect scattering cross-section evolution. The identification of this decoupled degradation paradigm provides critical insights for radiation-hardened circuit co-design, enabling simultaneous optimization of threshold voltage stability and drain current retention through strategic balancing of oxide defect passivation techniques and electron mobility-enhancing channel engineering, ultimately advancing the development of dual stress-tolerant integrated circuits for extreme radiation environments.

Author Contributions: Conceptualization, R.G.; validation, T.M.; formal analysis, X.Z. (Xiaowen Zhang); data curation, X.Z. (Xianyu Zhang); writing—original draft, Y.H.; writing—review and editing, Y.Y. All authors have read and agreed to the published version of the manuscript.

Funding: This research was funded by Guangdong S&T Program (grant No. 2022B0701180002) and Young Scientists Fund of the National Natural Science Foundation of China (Grant No. 62204061).

Data Availability Statement: The original contributions presented in this study are included in the article. Further inquiries can be directed to the corresponding author.

Conflicts of Interest: Authors Yujuan He, Rui Gao, Teng Ma, Xiaowen Zhang, and Xianyu Zhang were employed by the company Science and Technology on Reliability Physics and Application of Electronic Component Laboratory, China Electronic Product Reliability and Environmental Testing Research Institute. The remaining authors declare that the research was conducted in the absence of any commercial or financial relationships that could be construed as a potential conflict of interest.

We declare that we have no known competing financial interests or personal relationships that could have appeared to influence the work reported in this paper entitled "Total Ionizing Dose Effects on Lifetime of NMOSFETs Due to Hot-Carrier-Induced Stress".

References

1. Oldham, T.R.; McLean, F.B. Total ionizing dose effects in MOS oxides and devices. *IEEE Trans. Nucl. Sci.* **2003**, *50*, 483–499. [CrossRef]
2. Zheng, Z.; Zhao, X.; Zhao, K.; Gao, J.; Li, B.; Yu, F.; Li, B.; Luo, J.; Han, Z.; Liu, X. Comparison of the Total Dose Responses of Fully Depleted SOI nMOSFETs With Different Geometries for the Worst Case Bias Conditions. *IEEE Trans. Nucl. Sci.* **2019**, *66*, 2207–2214. [CrossRef]
3. Fleetwood, D.M.; Xiong, H.D.; Lu, Z.Y.; Nicklaw, C.J.; Felix, J.A.; Schrimpf, R.D.; Pantelides, S.T. Unified model of hole trapping, 1/f noise, and thermally stimulated current in MOS devices. *IEEE Trans. Nucl. Sci.* **2002**, *49*, 2674–2683. [CrossRef]
4. Chenming, H.; Simon, C.T.; Fu-Chieh, H.; Ping-Keung, K.; Tung-Yi, C.; Terrill, K.W. Hot-electron-induced MOSFET degradation—Model, monitor, and improvement. *IEEE Trans. Electron Devices* **1985**, *32*, 375–385. [CrossRef]
5. Heremans, P.; Bellens, R.; Groeseneken, G.; Maes, H.E. Consistent model for the hot-carrier degradation in n-channel and p-channel MOSFETs. *IEEE Trans. Electron Devices* **1988**, *35*, 2194–2209. [CrossRef]
6. Cui, J.-W.; Yu, X.-F.; Ren, D.-Y.; Lu, J. The influence of channel size on total dose irradiation and hot-carrier effects of sub-micro NMOSFET. *Acta Phys. Sin.* **2012**, *61*, 026102. [CrossRef]
7. Silvestri, M.; Gerardin, S.; Paccagnella, A.; Faccio, F. Degradation Induced by X-Ray Irradiation and Channel Hot Carrier Stresses in 130-nm NMOSFETs With Enclosed Layout. *IEEE Trans. Nucl. Sci.* **2008**, *55*, 3216–3223. [CrossRef]
8. Arora, R.; Fleetwood, Z.E.; Zhang, E.X.; Lourenco, N.E.; Cressler, J.D.; Fleetwood, D.M.; Schrimpf, R.D.; Sutton, A.K.; Freeman, G.; Greene, B. Impact of Technology Scaling in sub-100 nm nMOSFETs on Total-Dose Radiation Response and Hot-Carrier Reliability. *IEEE Trans. Nucl. Sci.* **2014**, *61*, 1426–1432. [CrossRef]
9. Ren, D.-Y.; Yu, X.-F.; Erkin. The relationship between hot-carrier damage and ionizing radiation damage in MOS capacitors. *Res. Prog. SSE.* **2001**, *21*, 103–108. (In Chinese)
10. He, Y.-J.; Zhang, X.-W.; Liu, Y. Total Dose Dependence of Hot Carrier Injection Effect in the NMOS Devices. *Acta Phys. Sin.* **2016**, *65*, 246101. (In Chinese) [CrossRef]
11. Xie, R.; Qiao, M.; Zhou, X.; Ge, C.; Xu, S.; Hong, G. The Correlation of Total Dose Effect and Hot Carrier Injection for Radiation Hardened MOS. In Proceedings of the 2022 10th International Symposium on Next-Generation Electronics (ISNE), Wuxi, China, 22–23 October 2022. [CrossRef]
12. Yogi, P.; Kumar, M.; Aditya, K.; Gupta, C.; Dixit, A. Impact of Hot Carrier Injection on Total Ionizing Dose Effect of 10-nm N-channel Bulk FinFETs. In Proceedings of the 2020 5th IEEE International Conference on Emerging Electronics (ICEE), New Delhi, India, 26–28 November 2020; pp. 1–4.
13. Zheng, Q.; Cui, J.; Yu, X.; Li, Y.; Guo, Q. Anneal Behavior of Total Ionizing Dose Irradiated UTBB FD-SOI n-MOSFETs Activated by Hot Carrier Stress. *IEEE Trans. Nucl. Sci.* **2023**, *70*, 37–43. [CrossRef]
14. Yao, R.; Lu, H.; Zhang, Y.; Zhang, Y.; Qiao, J.; Sun, J.; Xun, M.; Yu, G. Influence of Hot Carrier Degradation on Total Ionizing Dose in Bulk I/O-FinFETs. *IEEE Trans. Device Mater. Reliab.* **2024**, *24*, 456–462. [CrossRef]
15. Ren, Z.; An, X.; Li, G.; Liu, J.; Xun, M.; Guo, Q.; Zhang, X.; Huang, R. TID Response and Radiation-Enhanced Hot-Carrier Degradation in 65-nm nMOSFETs: Concerns on the Layout-Dependent Effects. *IEEE Trans. Nucl. Sci.* **2021**, *68*, 1565–1570. [CrossRef]
16. *BS EN 62416:2010*; IEC 62416-2010 Standard: Semiconductor Devices-Hot Carrier Test on MOS Transistors. International Electrotechnical Commission (IEC): Geneva, Switzerland, 2010.
17. Sze, S.M.; Kwok, K.N. *Physics of Semiconductor Devices*; A JOHN WILEY & SONS, JNC. PUBLICATION: Hoboken, NJ, USA, 2006; pp. 293–373. [CrossRef]
18. McWhorter, P.J.; Winokur, P.S. Simple technique for separating the effects of interface traps and trapped-oxide charge in metal-oxide-semiconductor transistors. *Appl. Phys. Lett.* **1986**, *48*, 133–135. [CrossRef]

Disclaimer/Publisher's Note: The statements, opinions and data contained in all publications are solely those of the individual author(s) and contributor(s) and not of MDPI and/or the editor(s). MDPI and/or the editor(s) disclaim responsibility for any injury to people or property resulting from any ideas, methods, instructions or products referred to in the content.

electronics

MDPI

Article

Comprehensive Study of Proton and Heavy Ion-Induced Damages for Cascode GaN-Based HEMTs

Huixiang Huang [1], Zhipeng Wu [1,2], Chao Peng [2,*], Hanxin Shen [1], Xiaoqiang Wu [1], Jianqun Yang [3], Zhifeng Lei [2], Xiuhai Cui [3], Teng Ma [2], Zhangang Zhang [2], Yujuan He [2], Yiqiang Chen [2] and Guoguang Lu [2]

1 School of Optoelectronics and Communication Engineering, Xiamen University of Technology, Xiamen City 361024, China; 13489938776@163.com (Z.W.); 2010111001@xmut.edu.cn (H.S.); 13860370478@163.com (X.W.)

2 Science and Technology on Reliability Physics and Application of Electronic Component Laboratory, China Electronic Product Reliability and Environmental Testing Research Institute, Guangzhou 511370, China; leizhifeng@ceprei.com (Z.L.); mateng@ceprei.com (T.M.); zhangangzhang@163.com (Z.Z.); heyu0000@126.com (Y.H.); yiqiang-chen@hotmail.com (Y.C.); luguog@126.com (G.L.)

3 National Key Laboratory of Material Behavior and Evaluation Technology in Space Environment, Harbin Institute of Technology, Harbin 150001, China; yangjianqun@hit.edu.cn (J.Y.); cuixiuhai@hit.edu.cn (X.C.)

* Correspondence: hxhuang@xmut.edu.cn

Abstract: Proton and heavy ion irradiation experiments were carried out on Cascode GaN HEMT devices. Results show that device degradation from heavy ion irradiation is more significant than from proton irradiation. Under proton irradiation, obvious device degradation occurred. Low-frequency noise testing revealed a notable increase in internal defect density, reducing channel carrier concentration and mobility, and causing electrical performance degradation. Under heavy ion irradiation, devices suffered from single-event burnout (SEB) and exhibited increased leakage current. Failure analysis of post-irradiation devices showed that those with leakage current increase had conductive channels without morphological changes, while burned out devices showed obvious damage between the gate and drain regions. SRIM simulation indicated that ionization energy loss-induced electron–hole pairs and displacement damage from nuclear energy loss were the main causes of degradation. Sentaurus TCAD simulation of heavy ion irradiated GaN HEMT devices confirmed the mechanisms of leakage current increase and SEB.

Keywords: Cascode GaN HEMT; proton irradiation; heavy ion irradiation; displacement damage; single event burnout

1. Introduction

Gallium Nitride has several advantages compared to other semiconductor substrate, such as wide bandgap (3.39 eV), high electron mobility, high thermal conductivity, high breakdown voltage, high-temperature and high-pressure resistance, and strong radiation tolerance [1]. Based on these outstanding characteristics, Gallium Nitride based devices display great competitiveness in the field of high-frequency and high-power electronic devices [2].

In recent years, several studies have been performed on the impact of proton irradiation on GaN HEMT device performance and the single-event effects (SEEs) in GaN HEMT devices due to heavy ion irradiation. X. W. Hu et al. [3] reported that the drain saturation current of AlGaN/GaN HEMT decreased after 1.8 MeV proton irradiation when the total dose reached 1×10^{12} cm^{-2}. D. Keum et al. [4] studied 1 MeV and 2 MeV proton irradiated AlGaN/GaN HEMT, where the drain saturation current decreased and the threshold voltage showed positive drift when the total dose reached 5×10^{14} cm^{-2} under 1 MeV and 2 MeV proton irradiation. L. Lv et al. [5] found that after 70 keV proton irradiation (with a total dose of 10^{13} cm^{-2}), the maximum drain saturation current density and the maximum transconductance density of AlGaN/GaN MIS-HEMT decreased, and the threshold voltage increased. References [6–9] studied single-event effects in GaN HEMT using protons and reported that a significantly negative effect on the device occurs only when the total proton dose exceeds 10^{14} cm^{-2}. In references [10–12], heavy ion irradiation was used to study the DC and RF characteristics of the devices before and after irradiation, and the results indicated that heavy ion irradiation increased the gate leakage current and source-drain leakage current of the devices.

Currently, a substantial amount of research primarily focuses on depletion-mode GaN devices and p-type gate structure enhancement-mode GaN devices subject to low-energy protons or low-energy heavy ions. However, there is a lack of reported research on the effects of high-energy protons or high-energy heavy ions on Cascode GaN HEMTs, particularly the comparative studies on the impacts of high-energy protons and high-energy heavy ions on Cascode GaN HEMTs.

In this work, proton and heavy ion irradiations were both performed on Cascode GaN HEMTs. The variations in the radiation damage induced by the two kinds of radiation sources were comparatively studied. The mechanisms of different radiation damages were analyzed based on electrical characterization and failure analysis techniques. After heavy ion irradiation, the devices were positioned using Emission Microscopy Imaging (EMMI) to locate the burnout points, and the device's failure modes were observed using a Focused Ion Beam (FIB) microscope, and finally the Sentaurus TCAD simulation facilitating the explanations of the damage mechanism.

2. Samples and Experimental Setups

The devices under test (DUTs) selected in this work are normally-off cascade AlGaN/GaN HEMT devices (TP65H050WS, 650 V, 50 mΩ) from Transphorm company, Goleta, CA, USA. Figure 1 shows the Cascode device structure. It consists of a lower layer with a high-voltage depletion-mode GaN HEMT and an upper layer with a low-voltage enhancement-mode Si MOSFET, both of which are cascade-connected with a common source and gate. When the Si MOSFET is turning on, the drain-source voltage of the Si MOSFET is forward close to 0 V, at which point the gate-source voltage of the GaN HEMT is negatively close to 0 V and greater than its negative threshold voltage, enabling the GaN HEMT to conduct. When the Si MOSFET is turned off, the externally applied source-drain bias divides across the channels of the GaN HEMT and Si MOSFET, causing the normally-on GaN HEMT's gate-source voltage to be less than its threshold voltage, thus turning it off.

The proton and heavy ion irradiation experiments were both carried out at the Space Environment Simulation and Research Infrastructure (SESRI) at Harbin Institute of Technology. This facility encompasses multiple key technologies including vacuum, cryogenics, dust particles, electromagnetic radiation, electron/proton radiation, plasma sources, weak magnetic fields, in situ or semi in situ online measurement and analysis, and life behavioral characteristic analysis. The protons had an energy level of 300 MeV, a dose rate of 2×10^{8} cm$^{-2}\cdot$s^{-1}, and a beam spot area of 1.5×3 cm^{2}, resulting in a total dose of 1×10^{11} p/cm^{2}. Proton experiments were performed at room temperature with a high voltage of 650 V applied to the drain. The Keysight B1500A semiconductor parameter analyzer was employed to conduct comprehensive parameter characterization of the devices before and after irradiation. A theoretical analysis was performed on the post-irradiation variations in electrical characteristics. Subsequently, the low-frequency noise testing system was utilized to measure and analyze defects in the devices pre- and post-proton irradiation, thereby validating the conclusions derived from the semiconductor parameter analyzer. Finally, SRIM simulation software (V2013) is used to simulate the ionization and nuclear energy losses induced by 300 MeV proton irradiation into the devices with depth to analyze and verify the defect formation mechanism due to electron–hole pair excitation and displacement damage effects of proton irradiation on Si MOSFET and GaN HEMT devices.

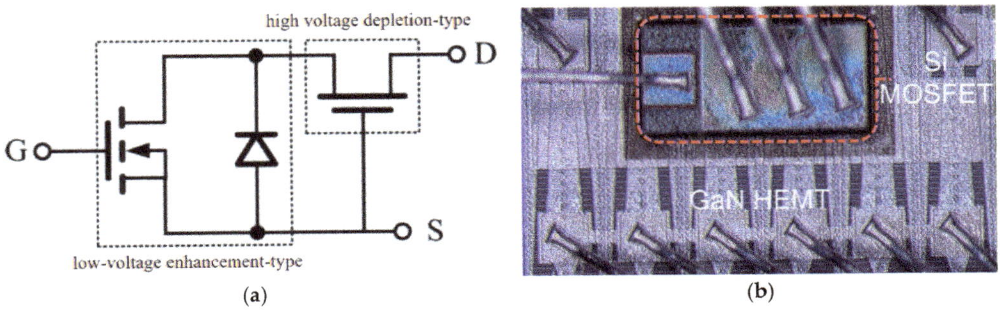

Figure 1. Cascode device structure. (**a**) Equivalent schematic; (**b**) physical diagram of the device.

The heavy ion irradiation used Kr ions with an energy of 500 MeV, a beam spot area of 1.5 × 3 cm^2, and a total dose of 3.167 × 10^5 p/cm^2, with a linear energy transfer (LET) of 37 MeV·cm^2/mg and a range of 110.61 um in GaN. The heavy ion experiment was conducted at room temperature, with three devices from the same batch subjected to different drain biases of 100 V, 150 V, and 200 V (#1 device with 100 V bias, #2 device with 150 V bias, and #3 device with 200 V bias). The output characteristics, transfer characteristics, and gate leakage current of the device were measured using the Keysight Company (Santa Rosa, CA, USA) B1500 semiconductor parameter analyzer before and after heavy ion irradiation, and the causes of the changes in electrical parameters caused by irradiation were analyzed.

After irradiation, the experimental devices underwent a decapsulation process. The decapsulated devices were used to locate the failure areas with an emission microscope (EMMI) (Advaced Company, Shanghai, China), and a focused ion beam (FIB) (ThermoFisher Company, Waltham, MA, USA) was used to spot cut and observe the device burnout. The SRIM simulation software was used to simulate the irradiation of 500 MeV Kr ions incident on Cascode GaN HEMT devices, clarifying the impacts of ionization and nuclear energy loss effects from heavy ion irradiation on the devices. Finally, the Sentaurus TCAD simulation software (V2019) was utilized to deeply reveal the microscopic physical mechanisms causing the increased leakage current and single-event burnout effect in Cascode GaN HEMT devices.

3. Results and Discussions

3.1. Degradation of Electrical Parameters of Devices Induced by Proton Irradiation

Electrical characterization of samples before and after proton irradiation reveals significant degradation in the electrical performance of the enhanced Cascode GaN HEMT devices. However, it is worth to note that the Cascode GaN HEMT device did not exhibit the Single Event Burnout (SEB) effect under 300 MeV proton irradiation with a drain bias of 650 V. Figure 2a depicts the output characteristics of the GaN HEMT device before and after irradiation, with configuration of Vgs = 4.0–4.4 V and a drain voltage swept of 0–4.0 V. A noticeable decrease in the saturated output current can be observed after irradiation. For instance, the maximum drain current of the device decreases by 14.23% after irradiation for the output curves at Vgs = 4.4 V. Figure 2b further illustrates the output resistance curves of the device before and after irradiation, indicating that the output resistance of the device does not change significantly after proton irradiation. Figure 2c,d show the transfer characteristics curves in linear and logarithmic scale of the device before and after irradiation, with a test condition of Vds = 0.05 V and Vgs = 0–7 V. The peak transconductance of the device decreases by 5.59% after irradiation, and the threshold voltage negatively shifts by 0.27 V. Figure 2e displays the Igs-Vgs characteristic curve before and after irradiation, showing minimal change in gate leakage current.

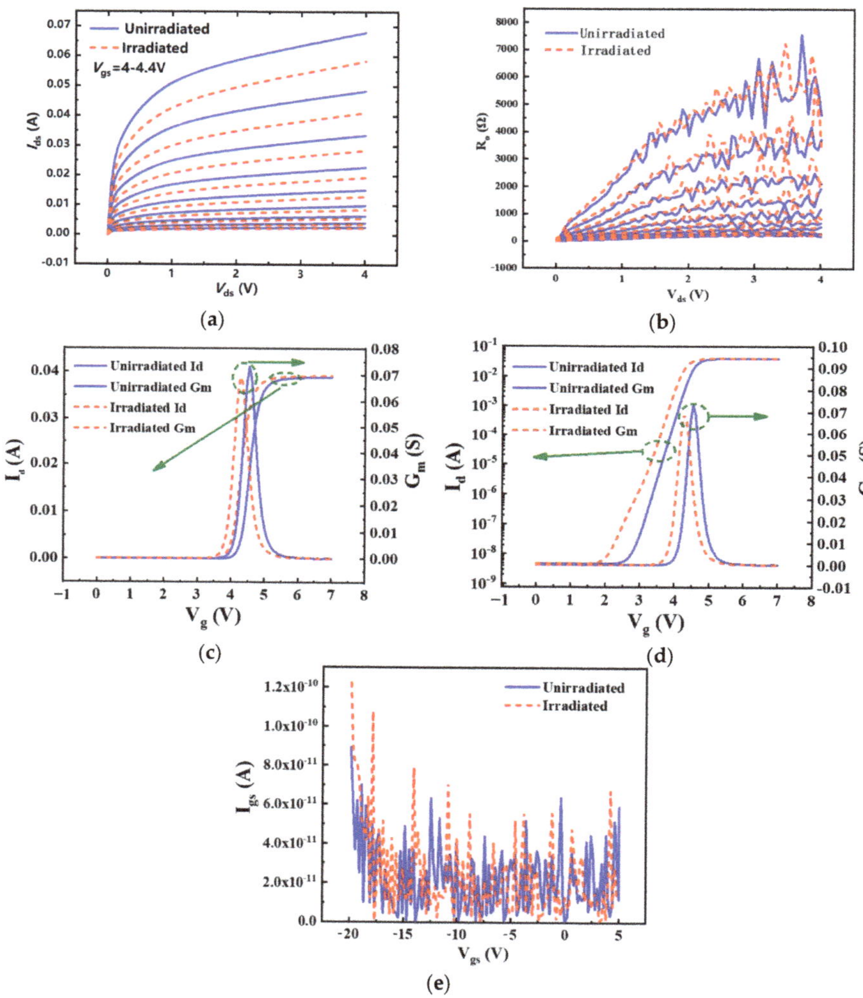

Figure 2. Electrical characteristic curves of Cascode GaN HEMTs under 300 MeV proton irradiation. (**a**) Output characteristics curve before and after irradiation; (**b**) the output resistance curve before and after irradiation; (**c**) the transfer characteristic curve before and after irradiation in linear scale and (**d**) the transfer characteristic curve before and after irradiation in logarithmic scale; (**e**) the I_{gs}-V_{gs} characteristic change curve before and after irradiation.

3.2. Heavy Ion Irradiation Effects

Heavy ion irradiation was conducted under three different voltage conditions. Device #1 drain biased at 100 V, #2 biased at 150 V and #3 at 200 V during irradiation. After heavy ion irradiation, device #1 exhibited a 5-order-of-magnitude increase in leakage current, and single-event burnout (SEB) occurred in devices #2 and #3. The drain current evolution with an ion energy of 500 MeV at 100 V drain bias are shown in Figure 3.

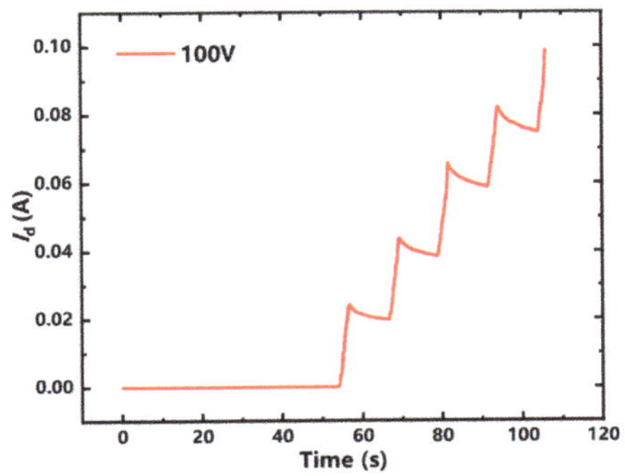

Figure 3. Monitoring of drain current during heavy ion Irradiation.

When the drain bias is set to 100 V, the drain current gradually increases with heavy ion irradiation. At a total dose of 3.167×10^5 p/cm^2, the current does not reach limit state, and the damage induced by the heavy ion irradiation to the device leads to a continuous increase in the drain current. Figure 4 presents the test results of device #1. In Figure 4a, the test conditions are Vgs = 4.05–4.4 V with a step size of 0.05 V, and the drain voltage varied from 0 to 4 V. Unlike proton irradiation, the output current significantly increases after irradiation. When the device is biased at Vgs = 4.05–4.4 V, the device operates in the subthreshold region. The increased output current at this bias suggests an introduction of additional leakage current paths after irradiation. Figure 4b represents the output resistance curve of the device before and after irradiation, indicating that, unlike proton irradiation, the output resistance of the device after heavy ion irradiation decreases and approaches 0, indicating serious electrical degradation or structural damage within the device. Under high-energy heavy ion irradiation, the device exhibits leakage-induced degradation. As leakage current rises with increasing irradiation dose, the device ultimately loses its reverse-blocking ability. Figure 4c,d show the transfer characteristics of the device before and after irradiation, with test conditions of Vds = 0.05 V and Vgs = 0–7 V. After irradiation, the peak transconductance of the device decreases by 70.17%, the transfer slope also decreases, and the threshold voltage negatively shifts by 0.11 V. From the logarithmic coordinate characteristics curve, it can be observed that the off-state leakage current of the device increases by five orders of magnitude after heavy ion irradiation. Figure 4e demonstrates that the reverse leakage current and forward leakage current of the device show no significant change after irradiation.

Considering the electrical characteristics before and after irradiation, the enhancement-type Cascode GaN HEMT device was more sensitive to heavy ion irradiation. Compared with the results of proton irradiation experiments shown in Figure 2, Kr ions also led to a more severe decrease in device transconductance. Moreover, the off-state leakage current of the device increased by five orders of magnitude after Kr ion irradiation.

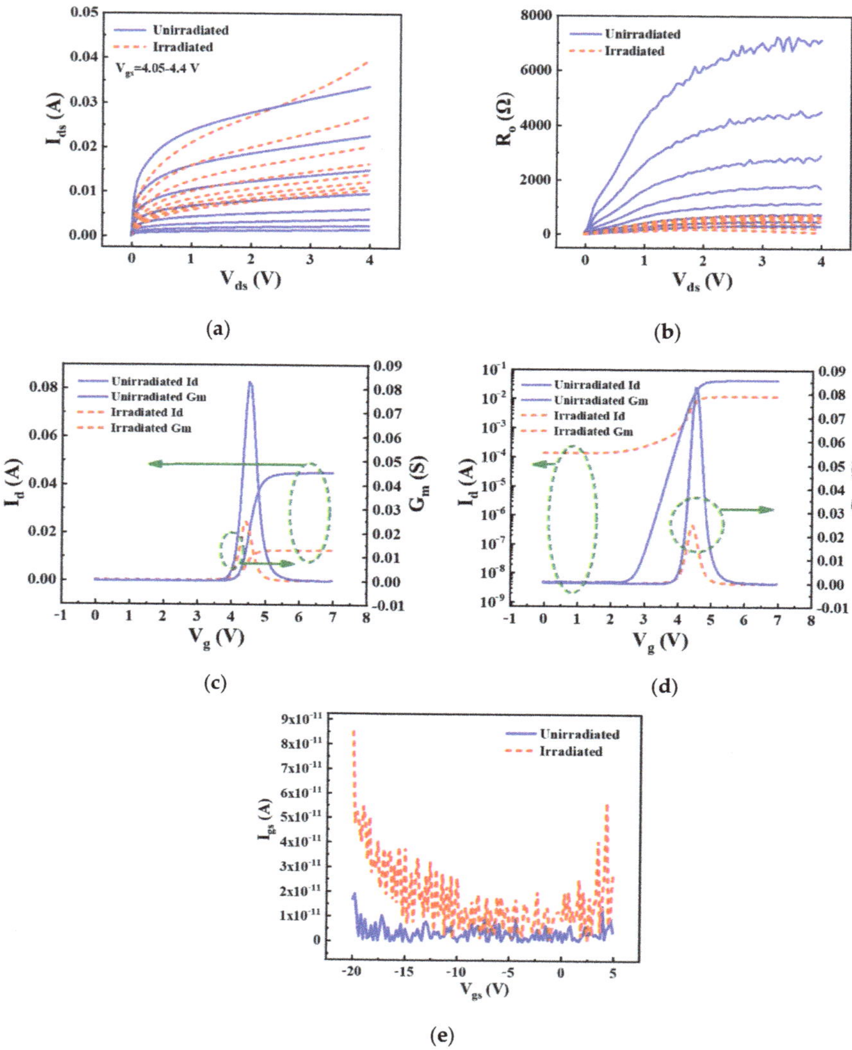

Figure 4. Electrical characterization curves of Cascode GaN HEMTs under 500 MeV heavy ion irradiation. (**a**) Output characteristics curve before and after irradiation; (**b**) the output resistance curve before and after irradiation; (**c**,**d**) the transfer characteristic curve before and after irradiation in linear and logarithmic scale, respectively; (**e**) the I_{gs}-V_{gs} characteristic change curve before and after irradiation.

3.3. Analysis and Discussions

3.3.1. Characterization of Low-Frequency Noise Defects in Devices After Proton Irradiation

Through comparison and analysis of the electrical characteristics of the device before and after proton irradiation, it can be inferred that the 300 MeV proton irradiation causes an increase in internal defects of the device. These radiation-induced defects serve as generation–recombination or trapping centers for a large number of carriers, resulting in a decrease in carrier concentration and carrier mobility at the channel of the device. This ultimately leads to the degradation of electrical characteristics such as saturation output current, threshold voltage, and transconductance.

In order to further characterize the changes in defects within the GaN HEMT device before and after 300 MeV proton irradiation, the flat-band voltage noise power spectrum density (S_{Vfb}) is used to assess the device's internal defects before and after the proton irradiation [13,14]. The relationship between the normalized channel current noise power spectrum density ($S_{id}/I_{ds}{}^2$) and the flat-band voltage noise power spectrum density can be expressed as follows [15]:

$$\frac{S_{id}}{I_{ds}^2} = \left(\frac{g_m}{I_{ds}}\right)^2 S_{Vfb}, \quad (1)$$

$$S_{Vfb} = \frac{qkTN_t}{fWLC_b^2(V_{gs} - V_{th})^2} \cdot \left(\frac{I_{ds}}{g_m}\right)^2. \quad (2)$$

The parameter g_m represents the transconductance of the AlGaN/GaN HEMT device, q is the elementary charge, k is Boltzmann's constant, T is the absolute temperature, and N_t is the volume trap density (cm^{-3}) in the gate oxide per eV. Additionally, f represents the frequency, W and L denote the width and length of the gate, respectively, C_b refers to the capacitance of the AlGaN barrier, and (V_{gs}-Vth) is the overdrive voltage. The magnitude of S_{Vfb} is entirely dependent on parameters such as trap charge and structural dimensions in the vicinity of the channel and is unrelated to the device's channel current and frequency [16–20]. Figure 4 illustrates the functional relationship between the noise power spectral density and the drain current before and after device irradiation. Based on Equation (1), the calculated values of S_{Vfb} for the device before and after proton irradiation are 3.2205×10^{-12} V^2·Hz^{-1} and 7.9659×10^{-11} V^2·Hz^{-1}, respectively. The relationship between defect density (N_t) and S_{Vfb} can be approximated as follows [19]:

$$N_t = \frac{WLC_b^2 f}{q^2 kT\lambda} S_{Vfb}, \quad (3)$$

λ is the conduction band alignment of the AlGaN/GaN heterojunction. As calculated in (3), after 300 MeV proton irradiation, the defect density N_t in the AlGaN/GaN device increased from 3.27×10^{16} cm^{-3}eV^{-1} to 7.17×10^{16} cm^{-3}eV^{-1}. This indicates an increase of 3.9×10^{16} cm^{-3}eV^{-1} in defect density within the device after 300 MeV proton irradiation. As can be seen from Figure 5, after proton irradiation, the defect density curve shifts upward, indicating an increase in the interface defects. This observation corroborates the aforementioned results.

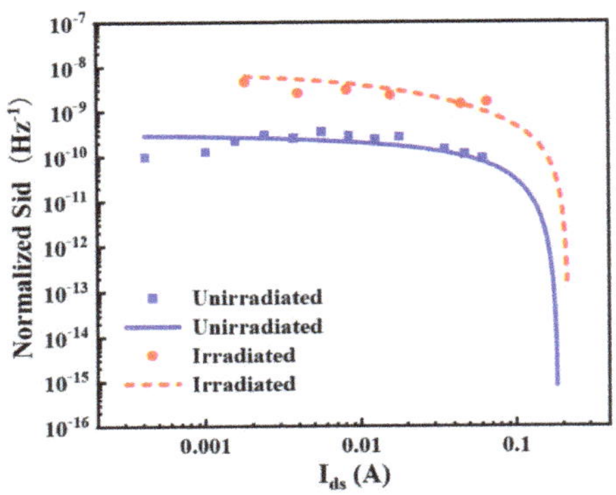

Figure 5. The variation in channel current noise power spectral density with channel current before and after the proton irradiation.

The low-frequency noise testing of Cascode GaN HEMT devices measures the gate oxide layer of the cascaded Si MOSFET. Thus, threshold voltage changes are mainly influenced by Si MOSFET performance variations. From the internal equivalent circuit diagram (Figure 1) of the Cascode structure, it can be seen that the most significant feature of this device is the cascading of low-voltage enhancement Si MOSFET devices, and the overall threshold voltage of the Cascode structure is mainly controlled by the Si-based MOS tube. The mathematical model of the MOSFET threshold voltage $Vth_{.Si}$ is [20]

$$V_{th.Si} = \frac{q}{C_{OX}}(N_{it} - N_{ot}) + \frac{\sqrt{4\varepsilon_{Si} q N_A \phi_{FP}}}{C_{OX}} + 2\phi_{FP} + \phi_{MS}, \tag{4}$$

$$\phi_{Fp} = \frac{kT}{q} In \frac{N_A}{N_i} \tag{5}$$

where N_{ot} is the oxide trap charge, N_{it} is the interface trap charge, C_{ox} is the unit area gate oxide capacitance, ε_{Si} is the dielectric constant of Si, N_A is the effective doping concentration in the P-type channel region, Ni is the intrinsic carrier concentration in silicon, Φ_{FP} is the Fermi potential of P-type substrate, and Φ_{MS} is the metal-semiconductor work function difference. The negative drift of Vth is mainly caused by electron–hole pairs induced by proton irradiation in the gate oxide layer. Due to the higher electron mobility in the oxide layer compared to holes, electrons can leave the gate oxide layer in a relatively short time. When most of the holes move toward the SiO_2/Si interface, they are captured by traps to form N_{ot} and N_{it}. Since oxide traps capture charges more easily than interface traps, N_{ot} increases much more than N_{it}, causing $Vth_{.Si}$ to decrease. Vth of the cascade device is directly related to the $Vth_{.Si}$ of the MOSFET device, so a decrease in Vth. When the Cascode device operates in the saturation region, the source-drain output current, I_{DS}, is mainly limited by the $I_{ds.Si}$ of the Si MOSFET, and thus I_{DS} is forced to limit as $I_{ds.Si}$ decreases [21].

The SRIM software was employed to simulate the ionization energy loss distribution of 300 MeV proton irradiation on Cascode GaN HEMT devices, which generates numerous electron–hole pairs. The nuclear energy loss distribution was also examined to reveal the displacement damage effects caused by proton irradiation, thereby explaining the proton irradiation damage mechanism. The simulation results are shown in Figure 6. Figure 6a presents the depth-dependent variations in ionization energy loss and nuclear energy loss in the enhancement-mode Si MOSFET device, while Figure 6b displays the corresponding depth-dependent variations in ionization energy loss and nuclear energy loss in the depletion-mode GaN HEMT device. The ionization energy of Si is 3.6 eV, and that of GaN is 8.9 eV, which is the energy needed to form an electron–hole pair upon irradiation. This suggests that GaN HEMT devices have better resistance to ionization-induced damage than Si MOSFET devices. As shown in Figure 6, after 300 MeV proton irradiation, ionization energy loss occurs in the devices. In the Si MOSFET, the ionization energy loss at the SiO_2/Si interface is 6.632×10^4 eV/Å, while in the GaN HEMT, the corresponding value at the AlGaN/GaN interface is 8.572×10^4 eV/Å. Analysis shows that proton irradiation causes more severe ionization damage in Si MOSFET devices. This generates a large number of electron–hole pairs, which in turn affect the electrical performance of the device.

It can also be seen from Figure 6 that nuclear energy loss occurs in the device under proton irradiation, indicating that displacement damage effects have taken place. The nuclear energy loss in Si MOSFET devices is higher than that in GaN HEMT devices. The displacement damage caused by nuclear energy loss creates numerous defects in the material, which ultimately leads to the degradation of the device's electrical properties. This is consistent with the above LFN experimental results. The SRIM software calculates that the energy required to form a vacancy in Si is 2.333 eV/Vacancy and in

GaN is 2.875 eV/Vacancy. This shows that under proton irradiation, Si MOSFET devices are more susceptible to displacement damage and trap charge formation. As a result, their post-irradiation degradation is more pronounced. Therefore, the degradation of Cascode GaN HEMT devices under proton irradiation is mainly due to the influence of Si MOSFET devices. Although GaN HEMTs are also affected by irradiation, the impact is relatively small.

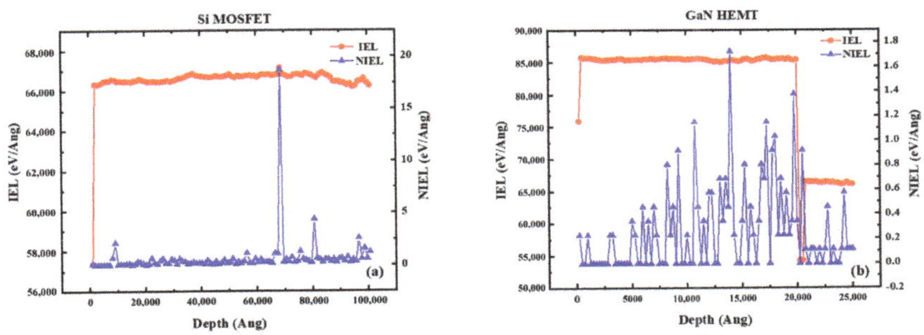

(a) (b)

Figure 6. Depth dependence of ionization and nuclear energy loss in the device. (**a**) Si MOSFET; (**b**) GaN HEMT.

3.3.2. Failure Analysis of Devices After Heavy Ion Irradiation

Compared to proton irradiation, Kr ion irradiation leads to a more severe reduction in transconductance. Furthermore, there is a five-orders-of-magnitude increase in off-state leakage current. From the analysis of the output characteristic curve in Figure 4a, it can be observed that the output current of the device decreased after proton irradiation compared to before irradiation, whereas the device irradiated with heavy ions exhibits an increasing trend. This is due to the increase in off-state leakage current that causes the output leakage current to rise in the heavy ion irradiated device when it is not fully turned on. That is, it indicates the presence of conductive channels between the source-drain regions of the device.

Considering that the metal wiring at the front side of GaN HEMT might cover the light emission, the EMMI analysis was conducted from the back side with the back metal stripped by grinding. The #1 device was opened, and an emission microscope (EMMI) was used to detect photons emitted and locate the fault points as shown in Figure 7. Multiple failure points distributed in different locations were found on the depletion GaN, while the enhancement Si MOSFET did not have any fault points. Each fault point represents a conductive channel and contributes to the increase in leakage current after heavy ion irradiation. The microscope was used to capture detailed images of the fault area of the device. It can be seen from Figure 7b that no significant material damage or morphological changes were observed at the failure location. Figure 8 presents the test results of the EMMI for device #2. It shows that the depletion-mode GaN HEMT device within the Cascode structure exhibits a Single Event Burnout (SEB) phenomenon, primarily occurring in the metal interconnect layer. The extracted details of the device fault points from Figure 8b reveal a significant burnout phenomenon between the gate and drain regions.

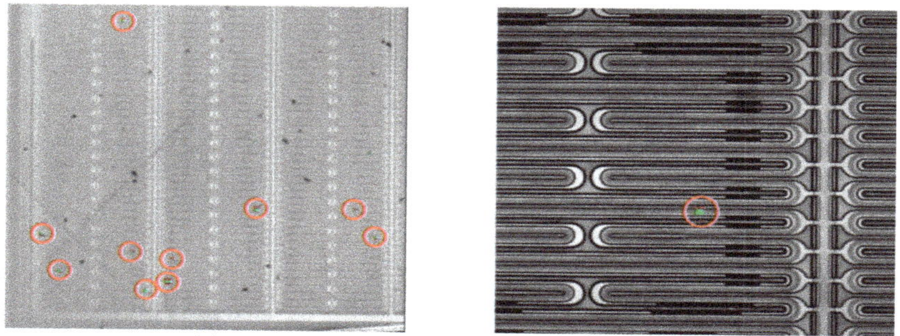

(**a**) (**b**)

Figure 7. (**a**) EMMI test results for device #1; (**b**) enlarge of a failure point. Failure points are emphasized by red circles.

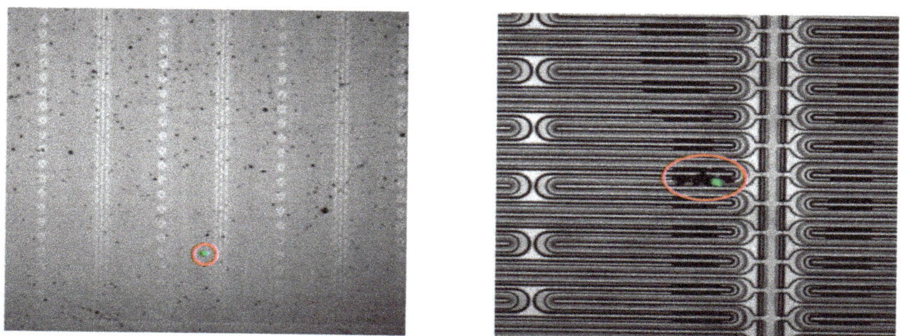

(**a**) (**b**)

Figure 8. (**a**) EMMI test results for device #2; (**b**) enlarge of the failure point. Failure points are emphasized by red circles.

Using EMMI for fault localization, targeted cutting of the burnt points was performed using focused ion beam microscopy (FIB) to observe the sectioned burnt points of the device. A comparison between Figure 9 indicates that there exists a void burnt between the gate and drain regions of the #2 device. The cross-sectional image cut by FIB in Figure 9c,d reveals that the internal metal of the device has been burned into a void due to the SEB effect.

Figure 10 illustrates the gate and drain current characteristics of the Cascode device as a function of drain voltage (V_d) at a gate voltage (V_g) of 4.2 V under off-state conditions after heavy ion irradiation. It can be observed that the gate current remains at the nA level before and after irradiation, indicating that the gate-source region is in the off state. However, the drain current increases with increasing drain voltage, suggesting the presence of a conductive channel between the source and drain regions. Further analysis suggests that the weakened Schottky barrier leads to a decrease in gate controllability, resulting in increased leakage current. When the leakage current becomes sufficiently large, a conductive channel forms between the source and drain.Based on the analysis of the internal circuit structure of the cascaded GaN device shown in Figure 1, it is evident that the TP65H050WS GaN device consists of a low-voltage enhancement-mode Si MOSFET and a high-voltage depletion-mode GaN HEMT connected in a common source and common gate configuration. When the device is in the reverse-biased cutoff state, the low-voltage enhancement-mode Si MOSFET is cutoff. The drain of the high-voltage depletion-mode GaN HEMT, the gate of the GaN HEMT, and the source of the Si MOSFET can form a leakage path between the source and drain, leading to high current phenomena at the source-drain terminals of the device. This is likely the primary reason for the SEB region in the high-voltage depletion-mode GaN HEMT observed in Figure 8 [22,23].

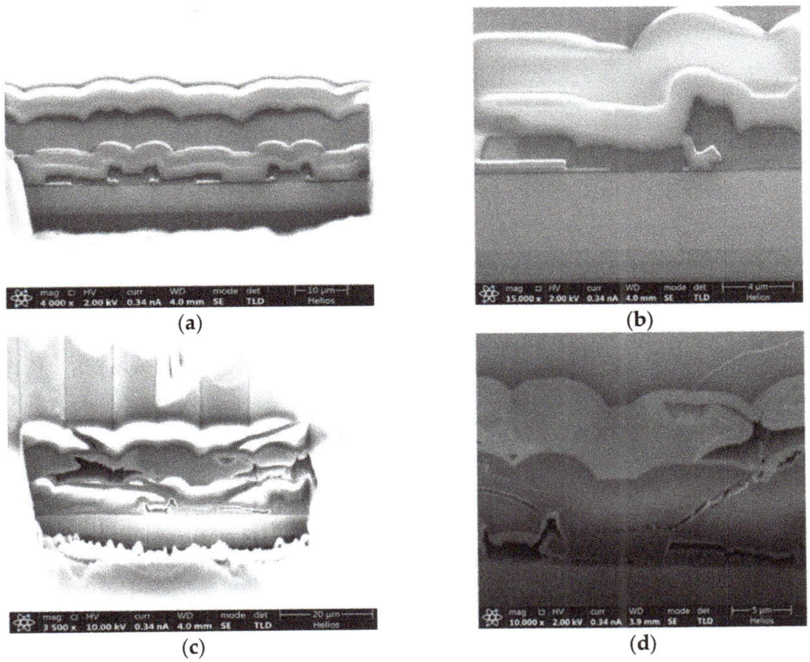

Figure 9. (**a**) Cross-sectional image of device #1 with off-state leakage current increase; (**b**) enlarge of the cross-section for device #1; (**c**) cross-sectional image of device #2 with SEB; (**d**) enlarge of the cross-section for device #2.

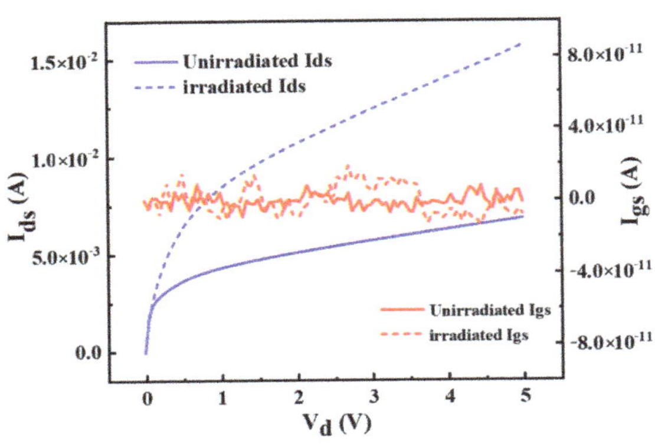

Figure 10. Variation in gate drain current with drain voltage of Cascode device in off-state mode after heavy ion irradiation.

To deeply explore the failure mechanisms of Cascode GaN HEMTs caused by Kr ions, 2D device simulations were conducted using the Sentaurus TCAD. The models used included carrier statistical, hot electron emission, heterojunction, carrier mobility, generation–recombination, piezoelectric polarization, tunneling, and heavy ion effects models. During simulation, the GaN HEMT was set to the off-state, with radiation particles vertically incident from the gate-drain region. To match the irradiation experiment, the ion

LET value was set to 0.543 pC/μm (about 37 MeV·cm^2/mg for GaN). The ion penetration depth was 3 μm to cover the entire device. The ion-induced failure factors in the device were examined by varying the drain bias [24].

Based on the aforementioned model, heavy ion irradiation simulation experiments were conducted on depletion-mode GaN HEMTs. The simulation structure diagram of the device is shown in Figure 11. From top to bottom: metal contact layer, silicon oxide passivation layer, AlGaN barrier layer, GaN channel layer, GaN or AlGaN buffer layer (Al composition of 0.02), AlN seed layer, and oxide layer. From left to right: source, gate, drain. Relevant parameters are listed in Table 1. Although discrepancies exist between the simulation device data/structure and the experimental device structure, the conclusions drawn through qualitative analysis suffice to provide theoretical guidance for this study.

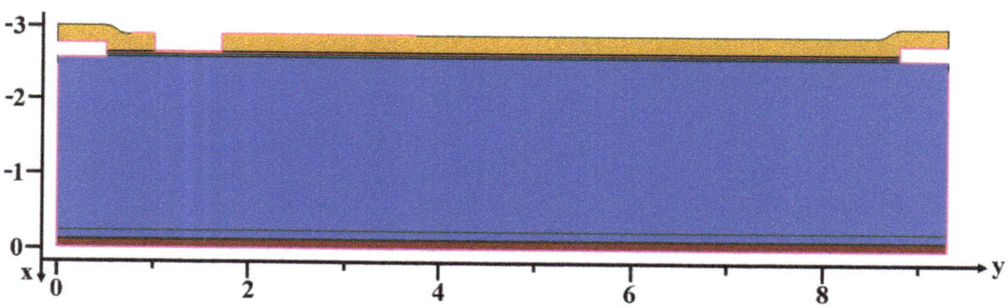

Figure 11. Depletion-mode GaN HEMT device simulation structure diagram. Gold: Silicon oxide passivation; Blue: AlGaN/GaN layer; Brown: oxide layer.

Table 1. Depletion-mode GaN HEMT device simulation structural parameters.

Parameters	Values
Source and drain length	0.5 μm
Gate Length	0.7 μm
Silicon oxide passivation	0.2 μm
AlGaN barrier layer	0.03 μm
GaN trench layer	0.03 μm
GaN or AlGaN buffer layer	2 μm
AlN seed layer	0.1 μm
Oxide layer	0.1 μm

Figure 12 presents the drain transient current variations in GaN HEMT devices caused by radiation ion incidence at different drain biases. As shown in Figure 12, at a drain bias of 100 V, the device's drain transient current variation is insignificant, and no SEB occurs. However, when the drain bias ranges from 150 V to 350 V, the drain current surges, exceeding 1 A. The resulting transient temperature rise causes thermal damage, leading to SEB in the device.

Figure 13 presents the gate transient current variations in GaN HEMT devices caused by radiation ion incidence at different drain biases. As shown in Figure 13, as the drain bias and particle irradiation time increase, the gate current rises and then drops to 0 A. This is because heavy ion incidence generates numerous electron–hole pairs. Electrons move towards the drain and are collected, while holes migrate towards the gate and accumulate. The accumulated holes in the AlGaN layer at the Schottky interface form a positive charge region, counteracting the electrons on the metal side and weakening the Schottky barrier.

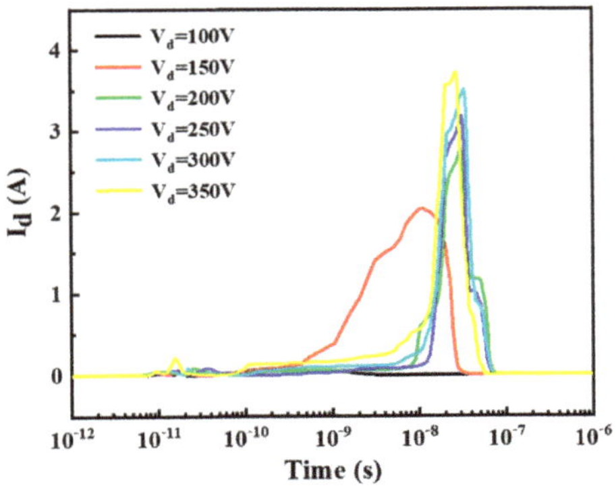

Figure 12. Drain transient current induced in GaN HEMT devices by ion incidence at different drain biases.

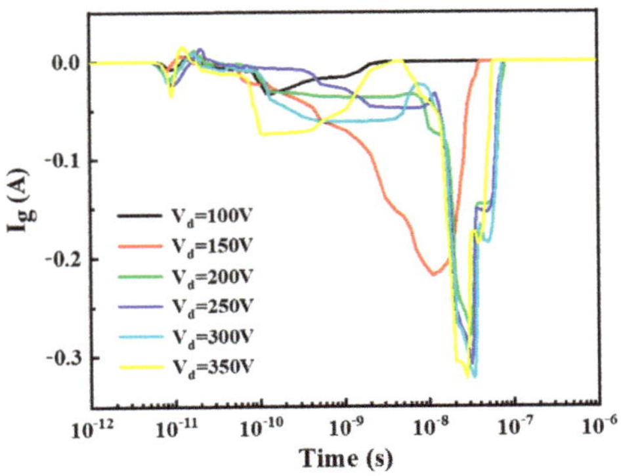

Figure 13. Gate transient current produced in GaN HEMT devices by ion incidence at different drain biases.

Figure 14 shows that regardless of the incidence angle of heavy ions, numerous holes gather below the gate. As the drain bias increases, more holes are driven towards the gate by the electric field, leading to hole accumulation at the gate. These holes are captured by defects at the interface caused by irradiation, forming a local positive charge layer. This partially offsets the intrinsic polarization-induced negative charge in the AlGaN layer, weakening the polarization electric field and narrowing the depletion region. The narrowed depletion region reduces the band bending of the Schottky barrier, lowering the barrier height. The reduced Schottky barrier allows more electrons to tunnel through, increasing leakage current. At higher drain biases, as shown in Figure 12, local transient large currents occur and persist for a while, gradually increasing the maximum lattice temperature within the device, causing thermal damage and SEB. As a conclusion, the degradation mechanisms observed in this study are compared to similar studies conducted on other semiconductor devices under radiation, as shown in Table 2.

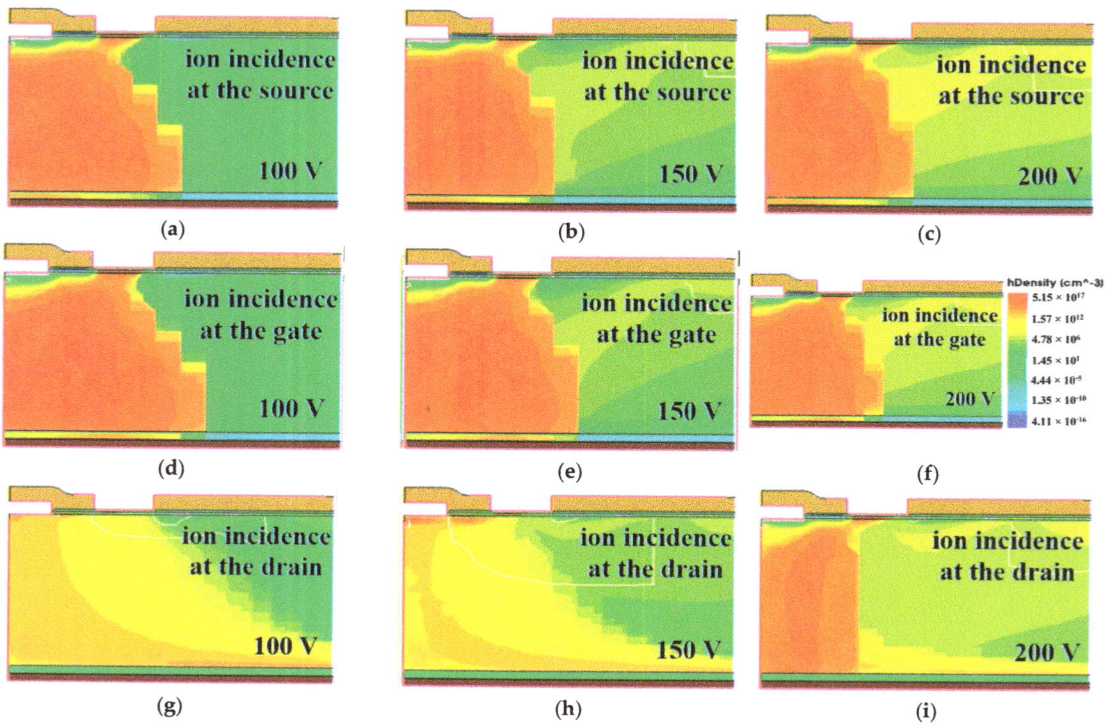

Figure 14. Hole distribution in GaN HEMT devices by ion incidence at different positions and drain biases. (**a**) source, 100 V; (**b**) source, 150 V; (**c**) source, 200 V; (**d**) gate, 100 V; (**e**) gate, 150 V; (**f**) gate, 200 V; (**g**) gate, 100 V; (**h**) gate, 150 V; (**i**) gate, 200 V.

Table 2. Degradation mechanisms comparisons to previous research.

Devices	Radiation Source	Failure Mode
SiC Schottky diode [25]	Ta, LET = 81.3 MeV·cm^2/mg	Single event burnout
65 nm N-channel MOSFET [26]	Sn ions	Negative drift of threshold voltage, increased leakage current
p-gate GaN HEMT [27]	Ge, LET = 37 MeV·cm^2/mg	Increased leakage current
Cascode GaN HEMT [27]	Ti, LET = 21.8 MeV·cm^2/mg	Single event burnout
Ga_2O_3 Schottky diode [28]	300 MeV proton	Single event burnout occurs at reverse 400 V.

4. Conclusions

Proton and heavy ion irradiation experiments were carried out on Cascode GaN HEMT devices. Results show that under 650 V drain bias, proton irradiation degrades the device's electrical characteristics without causing single-event burnout (SEB). Under 100 V drain bias, heavy ion irradiation significantly alters electrical properties, increasing the off-state drain current by five orders of magnitude. At 150 or 200 V drain bias, heavy ion irradiation induces SEB. Post-irradiation failure analysis and simulation results indicate that the electrical degradation after proton and heavy ion irradiation is due to the combined effects of ionization energy loss-generated electron–hole pairs and displacement damage caused by nuclear energy loss. The increased leakage current and SEB after heavy ion irradiation are attributed to a weakened Schottky barrier.

Author Contributions: Conceptualization, X.W. and J.Y.; methodology, Z.L.; formal analysis, Z.W.; investigation, C.P.; resources, X.C.; data curation, T.M.; writing—original draft preparation, Z.Z.; writing—review and editing, H.H. and H.S.; supervision, Y.H.; project administration, Y.C.; funding acquisition, G.L. All authors have read and agreed to the published version of the manuscript.

Funding: This research was funded by National Natural Science Foundation of China, grant number 12375268 and 12075065 and Foundation of National Key Laboratory of Materials Behavior and Evaluation Technology in Space Environment, grant number 6142910220209.

Data Availability Statement: The data presented in this study are available on request from the corresponding author due to privacy.

Conflicts of Interest: The authors declare no conflicts of interest. The funders had no role in the design of the study; in the collection, analyses, or interpretation of data; in the writing of the manuscript; or in the decision to publish the results.

References

1. Shen, L.; Heikman, S.; Moran, B.; Coffie, R.; Zhang, N.-Q.; Buttari, D.; Smorchkova, I.; Keller, S.; DenBaars, S.; Mishra, U. AlGaN/AlN/GaN high-power microwave HEMT. *IEEE Electron Device Lett.* **2001**, *22*, 457–459. [CrossRef]
2. Ambacher, O.; Smart, J.; Shealy, J.R.; Weimann, N.G.; Chu, K.; Murphy, M.; Schaff, W.J.; Eastman, L.F.; Dimitrov, R.; Wittmer, L.; et al. Two-dimensional electron gases induced by spontaneous and piezoelectric polarization charges in N-and Ga-face AlGaN/GaN heterostructures. *J. Appl. Phys.* **1999**, *85*, 3222–3233. [CrossRef]
3. Hu, X.; Choi, B.; Barnaby, H.; Fleetwood, D.; Schrimpf, R.; Lee, S.; Shojah-Ardalan, S.; Wilkins, R.; Mishra, U.; Dettmer, R. The energy dependence of proton-induced degradation in AlGaN/GaN high electron mobility transistors. *IEEE Trans. Nucl. Sci.* **2004**, *51*, 293–297. [CrossRef]
4. Keum, D.; Kim, H. Energy-dependent degradation characteristics of AlGaN/GaN MISHEMTs with 1, 1.5, and 2 MeV proton irradiation. *ECS J. Solid State Sci. Technol.* **2018**, *7*, Q159–Q163. [CrossRef]
5. Lv, L.; Lin, Z.; Guo, H.; Pan, X.; Yan, X. Proton irradiation effects on enhancement-mode AlGaN/GaN MIS-HEMTs devices. *Mod. Appl. Phys.* **2021**, *12*, 86–92.
6. Puzyrev, Y.S.; Roy, T.; Zhang, E.X.; Fleetwood, D.M.; Schrimpf, R.D.; Pantelides, S.T. Radiation-induced defect evolution and electrical degradation of AlGaN/GaN high-electron-mobility transistors. *IEEE Trans. Nucl. Sci.* **2011**, *58*, 2918–2924. [CrossRef]
7. Anderson, T.J.; Koehler, A.D.; Specht, P.; Weaver, B.D.; Greenlee, J.D.; Tadjer, M.J.; Hite, J.K.; A Mastro, M.; Porter, M.; Wade, M.; et al. Failure mechanisms in AlGaN/GaN HEMTs irradiated with 2 MeV protons. *ECS Trans.* **2015**, *66*, 15–20. [CrossRef]
8. Lv, L.; Ma, X.; Zhang, J.; Bi, Z.; Liu, L.; Shan, H.; Hao, Y. Proton irradiation effects on AlGaN/AlN/GaN heterojunctions. *IEEE Trans. Nucl. Sci.* **2015**, *62*, 300–305. [CrossRef]
9. Koehler, A.D.; Anderson, T.J.; Khachatrian, A.; Roche, N.J.-H.; Buchner, S.; Weaver, B.D.; Hobart, K.D.; Kub, F.J. Displacement damage and single event effects in AlGaN/GaN HEMTs. *ECS Trans.* **2017**, *75*, 13–20. [CrossRef]
10. Rostewitz, M.; Hirche, K.; Latti, J.; Jutzi, E. Single event effect analysis on DC and RF operated AlGaN/GaN HEMTs. *IEEE Trans. Nucl. Sci.* **2013**, *60*, 2525–2529. [CrossRef]
11. Poling, B.S.; Via, G.D.; Bole, K.D.; Johnson, E.E.; McDermott, J.M. Commercial-off the-shelf AlGaN/GaN HEMT device reliability study after exposure to heavy ion radiation. *Microelectron. Reliab.* **2017**, *68*, 13–20. [CrossRef]
12. Lei, Z.-F.; Guo, H.-X.; Zeng, C.; Chen, H.; Wang, Y.-S.; Zhang, Z.-G. Influence of heavy ion irradiation on DC and gate-lag performance of AlGaN/GaN HEMTs. *Chin. Phys. B* **2015**, *24*, 056103. [CrossRef]
13. Liu, Y.; Chen, H.B.; He, Y.J.; Wang, X.; Yue, L.; En, Y.F.; Liu, M.H. Radiation effects on the low frequency noise in partially depleted silicon on insulator transistors. *Acta Phys. Sin.* **2015**, *64*, 078501.
14. Christensson, S.; Lundström, I.; Svensson, C. Low frequency noise in MOS transistors-I theory. *Solid State Electron.* **1968**, *11*, 797–812. [CrossRef]
15. Chen, Y.Q.; Zhang, Y.C.; Liu, Y.; Liao, X.Y.; En, Y.F.; Fang, W.X.; Huang, Y. Effect of hydrogen on defects of AlGaN/GaN HEMTs characterized by low-frequency noise. *IEEE Trans. Electron Devices* **2018**, *65*, 1321–1326. [CrossRef]
16. Yue, S.; Lei, Z.; Peng, C.; Zhong, X.; Wang, J.; Zhang, Z.; En, Y.; Wang, Y.; Hu, L. High-fluence proton-induced degradation on AlGaN/GaN high-electron-mobility transistors. *IEEE Trans. Nucl. Sci.* **2020**, *67*, 1339–1344. [CrossRef]
17. Koehler, A.D.; Hobart, K.D.; Weatherford, T.R.; Kub, F.J.; Specht, P.; Anderson, T.J.; Weaver, B.D.; Greenlee, J.D.; Tadjer, M.J.; Porter, M.; et al. Proton radiation-induced void formation in Ni/Au-Gated AlGaN/GaN HEMTS. *IEEE Electron Device Lett.* **2014**, *35*, 1194–1196. [CrossRef]
18. Chen, Z.; Yue, S.; Wang, J.; Zhang, Z.; Huang, Y.; Wang, L.; Peng, C.; Zhong, X.; Lei, Z. Proton-induced effect on AlGaN/GaN HEMTs after hydrogen treatment. *IEEE Trans. Device Mater. Reliab.* **2021**, *21*, 297–302. [CrossRef]

19. Silvestri, M.; Uren, M.J.; Killat, N.; Marcon, D.; Kuball, M. Localization of off-stress-induced damage in AlGaN/GaN high electron mobility transistors by means of low frequency 1/f noise measurements. *IEEE Trans. Electron Devices* **2013**, *103*, 043506. [CrossRef]
20. Bai, R.; Guo, H.; Zhang, H.; Wang, D.; Zhang, F.; Pan, X.; Ma, W.; Hu, J.; Liu, Y.; Yang, Y.; et al. Study of high-energy proton radiation effects in GaN power devices with enhanced Cascode structure. *Acta Phys. Sin.* **2023**, *72*, 63–69. [CrossRef]
21. Huang, Y. Investigation of the mechanism of weak radiation-induced failure in GaN-based Cascode-type devices. *China Integr. Circuits* **2023**, *32*, 62–66.
22. Chen, R.; Liang, Y.; Han, J.; Wang, X.; Yang, H.; Chen, Q.; Yuan, R.; Ma, Y.; Shangguan, S. Experimental study on single-event and total-dose effects of gallium nitride-based high electron mobility transistors. *Acta Phys. Sin.* **2021**, *70*, 252–259.
23. Wan, P.; Yang, J.; Ying, T.; Lv, G.; Lva, L.; Dong, S.; Dongb, L.; Yu, X.; Zhen, Z.; Li, W.; et al. Effects of Ionization and Displacement Damage in AlGaN/GaN HEMTs Devices Caused by Various Heavy Ions. *IEEE Trans. Nucl. Sci.* **2021**, *68*, 1265–1271. [CrossRef]
24. Peng, C.; Lei, Z.; Zhang, Z.; He, Y.; Chen, Y.; Lu, G.; Huang, Y. Mechanism of SiC Schottky barrier diode damage due to heavy ion irradiation. *Acta Phys. Sin.* **2022**, *71*, 176101. [CrossRef]
25. Witulski, A.F.; Arslanbekov, R.; Raman, A.; Schrimpf, R.D.; Sternberg, A.L.; Galloway, K.F.; Javanainen, A.; Grider, D.; Lichtenwalner, D.J.; Hull, B. Single-Event Burnout of SiC Junction Barrier Schottky Diode High-Voltage Power Devices. *IEEE Trans. Nucl. Sci.* **2018**, *65*, 256–261. [CrossRef]
26. Ma, Y.; Gao, B.; Gong, M.; Willis, M.; Yang, Z.; Guan, M.; Li, Y. High fluence swift heavy ion structure modification of the SiO_2/Si interface and gate insulator in 65nm MOSFETs. *Nucl. Instrum. Methods Phys. Res. Sect. B Beam Interact. Mater. At.* **2017**, *396*, 56–60. [CrossRef]
27. Chen, R.; Liang, Y.-N.; Han, J.-W.; Wang, X.; Yang, H.; Chen, Q.; Yuan, R.-J.; Ma, Y.-Q.; Shangguan, S.-P. Single event effect and total dose effect of GaN high electron mobility transistor using heavy ions and gamma rays. *Acta Phys. Sin.* **2021**, *70*, 116102-1–116102-8. [CrossRef]
28. Li, X.; Jiang, W.; Wang, Y.; Zhang, H.; Peng, C.; Zhang, X.; Liang, X.; Fu, W.; Zhang, Z.; Lei, Z.; et al. Single-event burnout in β-Ga_2O_3 Schottky barrier diode induced by high-energy proton. *Appl. Phys. Lett.* **2024**, *125*, 092101. [CrossRef]

Disclaimer/Publisher's Note: The statements, opinions and data contained in all publications are solely those of the individual author(s) and contributor(s) and not of MDPI and/or the editor(s). MDPI and/or the editor(s) disclaim responsibility for any injury to people or property resulting from any ideas, methods, instructions or products referred to in the content.

MDPI AG
Grosspeteranlage 5
4052 Basel
Switzerland
Tel.: +41 61 683 77 34

Electronics Editorial Office
E-mail: electronics@mdpi.com
www.mdpi.com/journal/electronics

Disclaimer/Publisher's Note: The title and front matter of this reprint are at the discretion of the Guest Editors. The publisher is not responsible for their content or any associated concerns. The statements, opinions and data contained in all individual articles are solely those of the individual Editors and contributors and not of MDPI. MDPI disclaims responsibility for any injury to people or property resulting from any ideas, methods, instructions or products referred to in the content.